CUTTING GOD IN HALF

— AND PUTTING THE PIECES
TOGETHER AGAIN

A NEW APPROACH TO PHILOSOPHY

NICHOLAS MAXWELL

PENTIRE PRESS

Pentire Press

Pentire Press
13 Tavistock Terrace
London N19 4BZ
editor@pentirepress.plus.com

First published 2010

A catalogue record for this book
is available from the British Library

ISBN   978-0-9552240-2-7

Printed by Lightning Source

CONTENTS

Diagrams

To my dear wife Christine van Meeteren

PREFACE

God, according to Christianity, Islam and Judaism, is a Being who is all-powerful, all-knowing and all-loving. Such a God is knowingly responsible for all human suffering and death brought about by natural causes (and even brought about by people since natural causes are always implicated). How can such a God be all-loving? In order to solve this problem, I argue, we need to sever the God-of-Cosmic-Power from the God-of-Cosmic-Value. The former is Einstein's God, the underlying unity in the physical universe responsible for all that occurs. Because it is impersonal, it can be forgiven the terrible things it does. The latter is what is of most value associated with conscious human life – and sentient life more generally.

Having cut God in half in this way, the problem then becomes to put the two halves together again – to see how the God-of-Cosmic-Value can exist and flourish embedded in the God-of-Cosmic-Power. This, I argue, is our fundamental problem – our fundamental *philosophical* problem, our fundamental *theoretical* problem of knowledge and understanding, and our fundamental *practical* problem of living (personal, social and global). It is, at root, a *religious* problem, and ought to be the central concern of academic inquiry and education, and indeed of all of life. Unfortunately, at present, it is not, in part because of our long-standing failure to cut God decisively in half, and thus appreciate the fundamental character of the problem that results.

I go on to indicate how, in outline, this fundamental religious problem can be solved. Theoretical physics, properly understood, seeks to depict the nature of the God-of-Cosmic-Power. But physics depicts only a highly selected aspect of all that exists. It leaves out of account the world we see, hear, touch and are a part of, the experiential world imbued with meaning and value – the God-of-Cosmic-Value, in other words. We can, in this way, see how the God-of-Cosmic-Value can exist embedded in the physical universe, the God-of-Cosmic-Power. Furthermore, we can see how *we* can exist, conscious beings of value, embedded in the physical universe. We can see how we can act with some measure of free will embedded as we are in the physical universe.

Darwinian evolution can be re-interpreted to make intelligible the process of the gradual evolution of life of value in the impersonal physical universe.

But if we are to do better than at present at helping the God-of-Cosmic-Value to *flourish* within the God-of-Cosmic-Power, we need to *learn* how to do it, which means in turn that we have in our possession institutions of inquiry and learning rationally devoted to this task. It is just this that we do not have at present. For both intellectual and humanitarian reasons, we need to bring about a revolution in science, and in academic inquiry more generally, so that the basic task becomes to help life of value to flourish in the physical universe. The basic intellectual aim of academia needs to be, not knowledge, but rather *wisdom* – wisdom being the capacity to realize what is of value in life for oneself and others, thus including knowledge and technological know-how, but much else besides. I spell out arguments in support of this much needed revolution, and indicate what its implications would be, for natural science, for social inquiry, for the humanities, and for the structure and character of academia as a whole.

I go on to explain how this new kind of "wisdom-inquiry" would help us tackle our immense current global problems in rather more effective and cooperatively rational ways than we do at present.

The argument I develop has, in short, profound implications for science, for academic inquiry more generally, for education, for philosophy – indeed for our whole culture and way of life, the way we think about our problems, the world and our place in it.

In this book my concern is to indicate the path along which believers, especially believers in Christianity, Islam and Judaism, need to travel if they are to acquire a little more intellectual and moral integrity – *religious* integrity one might say. But even more important, perhaps, I seek to show how non-believers, agnostics and atheists, need to retrace the steps along the path that has led to their current position, to recover and develop much of value discarded by too hasty a past jettisoning of belief in God. This is not a book about the death of God. It is a book about how to improve our ideas about the nature of God. Believers and non-believers alike ought to pay attention. Both Richard Dawkins (author of *The God Delusion*) and Alister McGrath (co-author of *The Dawkins Delusion?*) should take note.

European culture – and thus, in a sense, world culture – has suffered a past gigantic rupture. Once upon a time everyone believed in God. Then we had, in succession, the Renaissance, the seventeenth century scientific revolution, the eighteenth century Enlightenment, the industrial revolution and the Darwinian revolution. Belief in God decayed; ceremonies associated with belief in God dwindled. The rupture took the form: "Once we believed in God; now we don't, but in some circumstances we observe ancient rituals and pretend that we do". Many, of course, resist this general decay of belief in God. Religious fundamentalists even try to turn their back on the modern secular world. But for many others, especially in those parts of the world most influenced by European culture, belief in God has been replaced by belief in science, in humanism, in liberalism, in democracy, socialism, freedom, progress, or the market – although, it has to be said, these latter beliefs are all looking, these days, a bit tarnished.

This rupture in European and world culture – from a God-dominated to a multi-faceted secular world (containing pockets of religious fanaticism) – vital and tremendous as it is in all sorts of ways, has nevertheless failed to develop ideas and values in the best possible way. As a result of rejecting God, instead of performing the surgical operation recommended here of cutting God in half, we have failed to develop properly what we have inherited from the rupture, and this inheritance has failed to come to full fruition. Science, education, humanism, liberalism, democracy, the arts, the market: all these suffer. Our culture, our whole modern world, is damaged. Above all, we fail to get into proper focus our fundamental problem: How to put the pieces together again once God has been sliced into two. How to help that which is of most value to flourish embedded as it is in the physical universe.

What we need to do, in short, is not *lose* our faith, but *improve* our faith, develop a rational faith, and above all try to put our rational faith into that which does really exist or can exist, and is genuinely of value. When we discover that God, in the traditional sense, does not and cannot exist, we need to work out carefully and delicately how our deepest aspirations, previously associated with the non-existent traditional God, can be developed in the best possible way, doing justice to the new universe we find ourselves in, and the new possibilities for what is of most value in that universe. The discovery of the non-existence of the traditional God impacts on our deepest, most personal desires, hopes and fears; and it impacts on the broadest, most public aspects and structures of our culture and society. Great care and sensitivity are needed to keep these threads in touch with one another, so that we may see how the deeply personal and the objectively social may be kept in touch with one another, so that both can develop in the best ways possible.

If our current ideals – science, humanism, liberalism, democracy, socialism, freedom, progress and the market – all seem these days somewhat tarnished, here is the reason: we have failed to perform the delicate operation of cutting God into two halves properly, and consequently have failed to get into focus properly what needs to be done to try to put the two halves together again. The secular "gods" that we have acquired as a result of the great rupture – science, humanism, etc. – have all emerged in crippled, distorted forms, in forms which fail to help what is of most value in life to flourish.

We need a religious revival – a religious revolution. We need to acknowledge and do justice to a religious dimension inherent in all our endeavours – political, educational, scientific, academic, even agricultural, industrial and commercial. But this needs to take the form of religious faith which meets elementary requirements of intellectual integrity and rationality, religious faith which sees the need to cut God in half, and which seeks to come to grips with the fundamental problem that results of putting the pieces together again, so that the God-of-Cosmic-Value is helped to flourish within the God-of-Cosmic-Power. Traditional religions and our current secular world fail to meet this challenge.

That is the line of argument I seek to trace out in this book.

CHAPTER ONE [1]

CUTTING GOD IN HALF

There is an urgent need to cut God in half.

To make such a pronouncement, these days, is a risky business. One risks having a fatwa declared against one. Or one risks being arrested for causing offence to the religious. So, let me explain as quickly as I can what I mean by this outrageous pronouncement, in the hope that this will keep such disasters at bay.

Consider the thesis: The ultimate reality is God, a Being who created the world and everything in it, a Being who is all-powerful, all-knowing, and all-loving, the source of all value, a Being who cares, profoundly, for the salvation of our souls.

This is, I take it, a central tenet of Christianity, Judaism and Islam. It is not upheld by all religions; it is not, for example, a part of Buddhism. But it is believed to be true by millions, possibly billions, of people alive today. Many more pay lip service to the doctrine. Its influence is felt in many contexts: educational, political, legal, ceremonial. It even has a certain impact on war and terrorism.

Given all this, the thesis deserves to be taken seriously intellectually. We need to ask: What would it be to treat this doctrine in an intellectually responsible fashion, with a measure of intellectual honesty? This question has especial relevance to education. For in so far as there are religious schools which take the thesis seriously, educational authorities responsible for educational standards need to be sure that these schools treat the God-doctrine in an educationally responsible way, in a way that promotes education rather than mere indoctrination. But we can only know what it means for a school to treat the God-doctrine in an educationally responsible way if we know what it means to treat it in an *intellectually* responsible way.

In what follows I spell out a few elementary steps that need to be taken if one is to uphold the God-thesis with a modicum of intellectual honesty. These are steps that *all* theistic religions need to

---

[1] This chapter is a modified version of Maxwell (2002a).

1

take if they are to avoid charges of dogmatism, charlatanism, disreputably and immorally misleading the public.

The first step towards intellectual honesty that one needs to take is to note that this creed, like other substantial theses about the nature of things, needs to be treated as a *conjecture*, a *hypothesis*, which may, or may not, be true.

Two points deserve to be noted about this first, minimal step.

First, even here, many religious traditions paralyse rational thought by making it a sin to doubt the existence of God. The sin, on the contrary, is to make doubt a sin.

Second, it is important that one does not exaggerate the power of reason, and claim that, in order to be rational, a belief must be proven or justified. As Karl Popper tirelessly argued, even our best scientific theories cannot be verified or justified; they remain, for ever, conjectures which, at best, can be empirically falsified.[2] The result of giving exaggerated powers to reason is that it becomes reasonable to hold that reason has its limits, and all sorts of beliefs, including religious ones, are beyond the scope of reason, defy reason, and are legitimately held as articles of irrational faith. Interpret the powers of reason more modestly, as helping us to choose, fallibly, between rival *conjectures*, and no thesis, not even a religious one concerning the existence of God, lies beyond the reach of reason.

The next step, then, is to ask: Are there good grounds for preferring the God conjecture to rival conjectures – such as, for example, that no such Being as God exists? What problems does the God-thesis solve, what phenomena does it explain? Can it be refuted?

There can be no doubt at all that the conjecture that an all-powerful, all-knowing, all-loving God exists does solve problems, and does make many things intelligible. The dreadful, apparently unsolvable problem of death is solved at a stroke: such a God would arrange for us to survive death. The dreadful problem of the unspeakable suffering of this world, the awful waste of human potential, the numbing injustice of human life: all this will be put right after death, if an all-powerful, all-knowing, all-loving God exists. And if such a God exists, we have an immediate explanation

---

[2] See Popper (1963) or, at a more technical level, Popper (1959).

2

for the fact that our environment, here on earth seems, in many ways (if not in all ways) especially designed to nourish us and support our existence. We can even understand why the universe is knowable to us, by means of science: God created both the universe, and human beings; being benevolent, He would naturally arrange things in such a way that there is a sufficient match between the nature of the universe and the nature of our minds for us to be able to improve our knowledge of the universe. God explains why science works.

How Can God's Evil Deeds be Excused?
But there is a dreadful snag. If God is all-knowing and all-powerful, then God must be knowingly in charge of natural phenomena, in particular those natural phenomena that cause human suffering and death as a result of earthquakes, drought, disease, accident. Even when people torture and kill other people, God is always a co-torturer and co-murderer, in that He decides the knife will not, at the last minute, turn into rubber, the bullet will not evaporate before it hits its target, the virus will not die or become abruptly harmless. Day after day, hour after hour, God knowingly tortures and murders innocent children (children dying of painful diseases) – to put the point at its most emotionally inflammatory, but correctly.

The obvious conclusion to draw is that the hypothesis that an all-powerful, all-knowing, all-loving God exists is refuted by the most elementary tragic facts of human existence.[3] This conclusion is

---

[3] Strangely enough, Richard Dawkins (2006) is rather dismissive of this decisive reason for rejecting the hypothesis that the traditional God exists. "it is" Dawkins remarks "an argument only against the existence of a good God. Goodness is no part of the *definition* of the God hypothesis, merely a desirable add-on." And he goes on to remark "it is childishly easy to overcome the problem of evil. Simply postulate a nasty God" (Dawkins, 2006, p. 135). But no Christian or Muslim who believes in the traditional God can conceivably calmly acknowledge that God may not be so good after all, and carry on as before, as Christian or Muslim, believing in a nasty God. Goodness is not an optional add-on: it is an absolutely essential ingredient of the traditional God. The manifest monstrosity, on a cosmic scale, of an all-powerful, all-knowing God (should He exist) is a devastating and lethal objection to the traditional

3

inescapable once one child has suffered and died as a result of injury or disease – suffered and died as a result of the knowing actions of God (if He exists). A loving God would take care of His children in at least as humane a fashion as, let us say, a petty thief. No run-of-the-mill petty thief would torture his child to death over a period of days or months, a commonplace action for God (if He exists). God tortures and murders billions of people; indeed none of us escapes.

Nothing can excuse God for killing one child, let alone all of humanity, one after the other. And yet, over the centuries theologians, instead of emphasizing that the God conjecture is decisively refuted, have instead struggled to invent excuses for God's criminal acts. The excuses are dreadful, utterly immoral and hopeless, and yet they continue to be taken seriously today.[4]

"God must allow us to suffer and die, because He must allow us our freedom" runs one excuse. So, should we equally demand of human parents that if their child runs onto the road in front of an incoming lorry, they should not interfere, so that the child may have his freedom? "God is unknowable, and we human beings cannot know why God performs these monstrous acts" runs another. But nothing can excuse God murdering a child slowly and agonizingly by means of cancer, let us say. People living in the Soviet Union under Stalin are on record as endlessly excusing the frightful crimes of Stalin; these excuses are morally and intellectually dreadful

---

God conjecture. As Stendhal said "The only excuse for God is that he does not exist" (quoted in Hicks, 1985, p. xi).

[4] The endeavour of attempting to excuse God's criminality even has a name, coined by Leibniz: *Theodicy*. Rarely does one even find the problem stated correctly. It is usually stated as the problem of understanding how God, being infinitely good, can allow evil to occur, and not as the problem of how an infinitely good God could *himself perform* endlessly many monstrously evil acts, torturing and maiming millions (if not billions in that we all suffer to a greater or lesser extent from natural causes during our lives) and murdering billions, in that we all die from natural causes, even those killed by their fellow human beings. For exercises in Theodicy see: Hicks (1985); McCord (1999); van Inwagen (2006); Swinburne (2003). For a compilation of writings on "the problem of evil", from Plato via Medieval times to the 20th century see Larrimore (2001).

4

(however excusable in the circumstances): how can any excuse, whatever it might be, be any better for God's far more dreadful crimes? "God lets us suffer so that we may grow spiritually" runs a third excuse. Are child molesters to be excused on similar grounds? Can we be so sure that suffering ennobles? Would not this argument imply that we do a person a favour if we hurt him? "It is not God who does these dreadful things, but the Devil". If God is all-powerful and all-knowing, God has the power to stop the Devil; if He decides not to, then He is in part responsible for what goes on. "People suffer and die because of the sins of their ancestors." What an appallingly immoral argument! "God does not murder people; he acts as a surgeon, causing pain in order to cure: those who die live on in Heaven (at least those who deserve it do)." But a surgeon who caused unspeakable pain in a patient over weeks or months, without adequate explanation, and without anaesthetics, would be struck off the medical register, and would doubtless be prosecuted for assault to an extreme degree: even if God does cause us to suffer so that we may be released into the after-life, this might mean that God does not murder, but it does not remotely excuse His actions. (On these grounds, no true believer could be accused of murder either, of course!)

Religious communities should hang their heads in shame at producing such appalling, immoral arguments. Taking such arguments seriously, even if only to set about refuting them, is in itself to take part in a corporate dance of insanity.

Why, why has humanity, or so much of humanity, allowed itself to be so bamboozled? Because the need for God is so potent, the fear of His non-existence so terrible. God's criminality is excused for the same reason, essentially, that Stalin's criminality was excused: the consequences of acknowledging that the crimes are real are too dreadful to contemplate. And this is backed up, in both cases, by a system of "education" which prompts one to believe that it is not God's (or Stalin's) criminality that is at issue, but one's own - any hint of a suspicion that God (or Stalin) is a monster instantly demonstrating one's own dreadful disposition for sin. How justified God (or Stalin) would be in punishing such suspicion, and how merciful God (or Stalin) so often proves to be in not bringing down instant punishment on those who so sin.

Granted that the conjecture that an all-powerful, all-knowing, all-loving God exists has been refuted,[5] what do we put in its place? This is the really important question!

In order to answer it, the first point we need to note is that, as we have discovered that God in the orthodox sense does not, and cannot exist, we need to be more open-minded about what sort of entity God may be. The question of whether God exists or not can always be converted into a question, not about God's existence, but rather about what *sort* of entity God is. If we mean something very specific and

---

[5] It may be held that my very brief, cavalier discussion of "the problem of evil" hardly does justice to the issue, and can hardly be said to amount to a refutation of the God conjecture. I think exactly the opposite is the case. It is quite impossible for there to be a God who is all-powerful, all-knowing, and all-loving, and attempts to argue that this is possible are intellectually and morally disreputable. State clearly that the problem is the appalling, billionfold criminality of God, far outdoing the actions of a mere Hitler or Stalin, and it is quite clear that there can be no excuse, no justification. The suffering and death of one child from cancer, caused by God – not merely permitted to happen by God – is enough to demolish the idea that there could be an omnipotent, omniscient, all-loving God. It does not matter who does it: torture and murder is not compatible with loving.

I might add that my concern in this book is not so much to demonstrate decisively that the traditional God cannot exist, which strikes me as utterly obvious, but to show how the traditional God conjecture, suffering from this dreadful defect, can be improved so as to overcome the defect, become intellectually and morally acceptable, and even become genuinely fruitful for intellectual and practical purposes. It is a commonplace, in science, that theories can be improved. Einstein's theory of gravitation is a considerable improvement over Newton's. The quantum theory of the electromagnetic field is, in many respects, an improvement over James Clerk Maxwell's classical theory of the electromagnetic field. And so on. But in the field of humanities, philosophy, politics and religion, we do not seem yet to have developed an analogous tradition of improving theses and ideas. It is this that I am attempting to do in this book in the field of religion (and philosophy). This attempt is intimately connected with a general *methodology* for improving aspects of life, which I call *aim-oriented rationality*, and which I shall expound and argue for in chapter 6.

highly problematic by "God", then it is all too likely that God, in this sense, does not exist. But if we mean something highly unspecific and unproblematic by "God", it becomes much more likely that "God", in this sense, does exist. One way of posing the question is: What is the nature of that Entity which (a) preserves as much as possible about what is best in the orthodox notion of God, and (b) exists? Reformulated in this way, the question becomes, not "Does God exist?", but rather "What is the nature of God?" where God, in this sense, exists by definition as it were.

The next step, in answering the above question, is to track down what it is in the orthodox conception of "God" that we have been discussing that makes it so impossible for "God" in this sense, to exist.   We then need to broaden our conception of "God" appropriately, so that "God", in this sense, becomes at least a viable possibility, the "God"-thesis a viable conjecture.

It is not difficult to track down what it is about the orthodox conception of God that creates the difficulties we have been considering. These all come from the supposition that God is *both* all-powerful, and all-loving. This is what we need, of course, an all-powerful being who is also all-loving, so that everything that is most precious in existence will be effectively, lovingly taken care of. But in our world, this leads to the awful consequences that we have been considering.

One possibility, of course, is that God, far from being loving, is thoroughly evil.   But this does not seem to do justice to all the wonderful things that there are in existence. What is so confusing is that life is such a mixture of joy and horror, the extraordinary, the prosaic and the unspeakable.

Perhaps God is confused, schizophrenic even, a dreadful mixture of love and hate? But this does not seem to do justice to the majesty of the universe, its intricate splendour. Could this have been created by a neurotic?

One might take the thing further, by postulating two equally powerful gods, God and the Devil, one good, the other evil, locked in terrible combat, humanity somehow the field of battle.[6] But if this

---

[6] This seems to have been the view of Georges Bernanos, the novelist: see, for example, Bernanos (1948).

really were the case, there would be, one feels, more disruptive explosions in the natural world, as the two cosmic Beings fought out their mighty, eternal battle.

Another possibility, of course, is that God is all-loving, but lacks power. He sees the terrible things that go on, but is powerless to intervene. It is a version of this hypothesis that I wish to defend. As it stands, however, it is incomplete: nothing is said about the nature of that which *does* have power, which *is* the cause of natural phenomena, and thus the cause of so much of our suffering.

Bisecting the Deity

Here is how, in my view, the problem is to be solved. God must be cut in two. (At last I come to the proposition with which I began.) The God of *power* must be severed from the God of *love*, the God that is the source of all *value*. Or, if it seems just a little too brutal, too grandiose, to speak of cutting God in half, let us say, rather, that we need to cut the *concept* of God in half – a much more modest surgical deed.

The God of cosmic power is utterly impersonal. It is that impersonal *something*, whatever It may be, that exists everywhere, eternally and unchanging, throughout all phenomena, and determines (perhaps probabilistically) the way phenomena unfold. It is what theoretical physics seeks to discover. It is Einstein's "God", eternal, omnipresent, all-powerful, but utterly impersonal, an It, not a conscious Being. [7] It is that physical property of the fundamental physical entity, the fundamental physical field or whatever, that determines the way in which that which changes *does* change. It is what corresponds physically to the true unified theory of everything that physicists seek to discover. [8]

It is this cosmic It that is responsible for all our suffering. And precisely because It is an It, incapable of knowing and feeling, It can be forgiven the terrible things that It does. If It knew that the laws of

[7] See A. Einstein (1973, pp. 36-52).

[8] Strictly speaking, it is what corresponds physically to the true unified theory of everything (the God-of-Cosmic-Power) plus variable physical states of affairs which, together, at any instant, determine (perhaps probabilistically) subsequent physical states of affairs.

nature, working themselves out as usual, meant, in this particular case, horrible suffering and death from cancer for this child, agonizing burns for this person, burial in rubble for that person, the It would at once bend a law of nature here and there, so that these ghastly tragedies can be avoided. But this cosmic It has no mind, no understanding, no awareness: it goes blindly on Its way, incapable of knowing anything, and therefore can be forgiven.

But what of the other half of the traditional God, the God of value? This, I suggest, is what is best in us. It is that potentially or actually aware and loving self within us that sees, feels, knows and understands, at least partially, and either *does* intervene to prevent disaster, or is powerless to do so. The God of value is the soul of humanity, embedded in the physical universe, striving to protect, to care for, to love, but all too often, alas, powerless to prevent human suffering. (More generally, the God of value is that which is of most value, actually and potentially, in sentient life.)

What is usually characterized as belief in science and humanism is actually, when properly interpreted, what emerges as a result of a rational evolution of belief in God. The scientific view of the universe, and the humanistic faith that it is individual human beings that are of supreme value in existence amount, when taken together, to a profoundly religious view, one which does not have the awful intellectual and moral defects of orthodox Theism (whether Christian, Judaic or Islamic). Science + Humanism only denies the existence of God in a thoroughly disreputable sense of "God"; granted a somewhat more reputable sense of "God", Science + Humanism is a passionate affirmation of the existence of God. But, as we shall see, *reinterpreting* Science + Humanism in this religious way has consequences for the nature of both science and humanism – and a range of other aspects of our culture as well.

Cutting God in half may solve problems that haunt orthodox Theism, but it does so at the expense of creating an immense new problem. Having chopped God into two, into the God-of-Cosmic-Power and the God-of-Cosmic-Value, we are at once confronted by the problem: How are the two halves to be put together again? How is it possible for the God-of-Cosmic-Value to exist embedded in the God-of-Cosmic-Power – the physically comprehensible universe? *How can we understand our human world, embedded as it is within the physical universe, in such a way that justice is done to both the richness, meaning and value of human life on the one hand, and what modern science tells us about the physical universe on the other hand?*

This problem (created by cutting God in half) is, quite simply, the most general and fundamental problem confronting humanity. It is a *philosophical* problem – indeed, the fundamental problem of philosophy: How is it *possible* for our human world, imbued with sensory qualities, consciousness, free will, art, science, and much else of value, to exist embedded in the physical universe? (This embraces, as subordinate issues, the mind-body problem, the problem of free will, problems of knowledge, of perception, of the philosophy of science, of biology and evolution, even problems of moral and political philosophy, problems of language, culture, history, abstract entities, time, space and causation.) The above is also a fundamental problem of *knowledge* and *understanding* much more generally – the basic problem of science: What is the nature of the physical universe? How precisely do features of our human world, such as perceptual qualities, consciousness, and life more generally, fit into the physical universe? The problem can also be regarded as a fundamental problem of *living*, of *action*: How can we help what is of value in existence, actually and potentially, to flourish? What do we need to do, as individuals, so that what is of value to us may flourish? And what do we need to do, collectively, socially and politically, so that what is of value to people everywhere, to humanity, may flourish? The problem of fitting the God-of-Value into the God-of-Cosmic-Power (the underlying unified It of the physical universe) is not only a conceptual problem, a problem of knowledge and understanding; it is also a *practical* problem, the most general, fundamental practical problem that there is: to help the God-of-Value, what is of most value in us, to exist in the physical universe in ways that are less painful and constrained, more exuberant and joyful, more just, peaceful and noble, than at present. Once we recognize that the God-of-Value is what is of most value, actually and potentially, in us, it becomes our most profound religious obligation to help what is of value in us to flourish in the real world.

10

The outcome of treating the initial God-thesis with a modicum of intellectual honesty is that we are led straight to the most fundamental problems of knowledge, understanding and living that there are. The character of these fundamental problems of thought and life is brought sharply into focus. And as a result, much is changed. Academic inquiry is transformed. It becomes a fundamentally *religious* enterprise: to improve our knowledge and understanding of how the cosmic God-of-Value fits into the cosmic God-of-Power and, above all, to help the former to flourish within the tight embrace of the latter. Education is transformed. All education becomes religious in character. It has, as a basic task, to explore aspects of the fundamental problem: How can the God-of-Value fit into, and flourish within, the God-of-Power? Politics is transformed. It too becomes religious, in that it seeks to implement policies which help the God-of-Value to flourish inside the God-of-Power. Our lives are transformed. Personal life too becomes religious in that a basic task is to discover how we can help that part of the God-of-Value associated with our life to flourish in the cosmic God-of-Power. The task, of course, is somehow to get the God-of-Power so to act that the God-of-Value flourishes. Even theoretical physics is transformed, in that it becomes a religious quest, that part of science devoted to discovering the precise nature of the cosmic God-of-Power. The traditional division between the religious and the secular is annihilated. The secular is entirely engulfed by the religious.

These, at least, are some of the changes that would be brought about were we to take seriously and act on the implications of the simple idea that putting the two halves of God together again is indeed our fundamental problem of thought and life. Much is lost if we merely discard the God conjecture altogether. Believers have much to learn from bringing some intellectual integrity to religion and to ideas of God – and non-believers have much to learn from this as well. In the rest of this book I tackle the problems created by cutting God in half, and do my best to show just how fruitful, potentially, it is for both thought and life, to take these problems seriously, as urgent and fundamental – truly *religious* in character.[9]

---

[9] It may be objected that there is nothing unique about my proposed

11

At this point it may be objected: But why continue to talk of God at all? Is it not far better to get rid of God altogether, and put our faith, straightforwardly, in science and humanism unadorned with irrelevant theological trappings? And in any case is not all this stuff about chopping God in half very old news? Did not Friederich Nietzsche declare God to be dead long ago in the 19th century?[10] In cutting God in half, am I not merely repeating what Nietzsche and others did long ago, in killing God off? How, in any case, could God survive being brutally cut into two pieces in the way I have recommended?

God is too important a notion to discard. It is a focus for fundamental issues. What, ultimately, is the explanation for everything? What is the ultimate purpose of life? What is ultimately of value in existence? These are among the questions the God-hypothesis seeks to answer. And this answer – the idea of God – has had a profound, long-standing impact on our culture and social world. We should not merely discard the notion, declare the whole idea to be defunct, or God to be dead. Rather, in the face of the devastating objections to the traditional God-hypothesis, we should do what I have indicated: *improve* the thesis so that it overcomes these devastating objections (while retaining as much of what is of

---

solution to the so-called "problem of evil". I have already acknowledged this to be the case. One could imagine God is an all-powerful, all-knowing Being who is utterly monstrous. Or one could imagine God is an all-knowing, all-loving being who has lost control of his creation – the universe – and is thus very far from being omnipotent. What I claim for my proposed solution – the Bisected God – is that it uniquely (a) preserves more of what is of value in the traditional God than any rival solution, (b) is an intellectually and morally worthy notion, a *religiously* worthy notion – unlike the traditional notion, (c) is such that there are good grounds for holding that God, in this sense, the Bisected God, does exist, and (d) the thesis that God, in this sense, does exist, is potentially extraordinarily fruitful, for both thought and life. Much of the rest of this book is devoted to arguing for (c) and (d). A religious view based on accepting the Bisected God is, on these grounds, far more worthy, intellectually, morally and religiously, than the really very disreputable views of traditional Christianity, Islam or Judaism.

[10] Nietzsche (2006), section 125.

value in the traditional thesis as possible). This serves at least two purposes (there are others as we shall see).

First, it holds out the hope of keeping alive an awareness of ideas and problems at a fundamental level in our culture. As I have already indicated, and as I will argue in some detail throughout the book, abandoning – instead of *improving* – the God-hypothesis has had damaging consequences for a range of endeavours and institutions, from science and the humanities to education, ideas about what is of value in existence, and our capacity to solve global problems intelligently, humanely and effectively. If God had been cut decisively in half in the way I am recommending long ago in the 17th century, let us say, this might not automatically have cured these ills, but it would have helped.

Secondly, *improving* rather than abandoning the God-thesis provides believers with an open road along which they may travel, rather than leaving them stuck in a cul-de-sac. If, in our culture, there are clear indications as to how the God-hypothesis can be improved so that it overcomes the devastating objections it faces, and becomes intellectually and morally acceptable – even fruitful – this is something individuals and groups can avail themselves of to learn, to improve their religious ideas and lives. But if our culture does no more than confront one with the stark choice, "hold onto an intellectually and morally bankrupt idea of God, or abandon the whole idea of God altogether", the chances are that believers will opt for the former choice, since otherwise they must simply abandon their fundamental beliefs. As I have said, the believer is left stuck in a cul-de-sac.

There is another option. It is to cease to take God seriously, soften and sentimentalize Him, shroud Him in metaphor and double-speak, so that nothing that is said is to be taken at face value. As a result, religious belief is turned into something subjective and intangible, beyond the scope of reason and criticism. But this option is perhaps even more intellectually and morally disreputable than that of holding on to traditional Theism. Ultimate questions about the nature of the world and the purpose of life deserve to be treated with clarity and intellectual integrity. Doing that enhances the possibility of learning, of improving our ideas and even, perhaps, our lives.

13

This is sabotaged when clarity and transparent content are converted into metaphor and double-speak [11]

---

[11] For a devastating critique of this way of defending traditional religion by retreating into metaphor and doublespeak, see Bartley (1962).

CHAPTER TWO

FUNDAMENTALISM WITH A VENGEANCE

So, we have dared to take the terrible sword of reason to God, and with two or three swift strokes have sliced God into two halves: the God-of-Cosmic-Power, and the God-of-Cosmic-Value. The God-of-Cosmic-Power is that impersonal *something*, present everywhere at all times in an unchanging form, inherent in all phenomena, which determines (perhaps probabilistically) the way all events unfold. It is what physics seeks to discover in hunting for "the true physical theory of everything". The God-of-Cosmic-Value is what is of value in us, in our human world, and in the world of conscious and sentient life more generally, actually and potentially. It is suffused throughout the world that we experience, inherent in woods, fields and estuaries, our homes and neighbourhoods, the laughter of a child, kindness to strangers, times of happiness, friendship, intimacy and love.

And now we are confronted by the appalling problem: Having cut God into two, how do we put the two halves together again? How can the God-of-Cosmic-Value be inside the God-of-Cosmic-Power? How can our human world as we experience it, full of sound, colour, feel, taste and smell, imbued with sentience, consciousness, free will, meaning and value, containing everything we love and care for, this whole experienced world in which we live, somehow be embedded in, or be an integral part of, the physical universe? How can the God-of-Value exist and flourish within the God-of-Cosmic-Power?

The God-of-Cosmic-Value and the God-of-Cosmic-Power intersect in us. On the one hand, the God-of-Value exists primarily, but not exclusively, in us and through us – through our miraculous experiencing, conscious life here on earth. And on the other hand, we are utterly miniscule, even though complex and intricate, fragments of the vast God-of-Cosmic-Power. With the miniscule snippet of Cosmic Power apportioned to us, to our physical bodies and brains, we must somehow help the God-of-Value to exist and grow inside the God-of-Cosmic-Power. We are burdened with the appalling responsibility of trying to help the God-of-Value to flourish inside the God-of-Cosmic-Power. This is our fundamental religious problem – the fundamental problem, indeed, that confronts us in life. It is, first, our fundamental *practical* problem of living – confronting each one of us individually as we live, and confronting humanity as a whole. How can we realize what is of most value to us, actually and potentially, in the real world, as we live? It is, second, our fundamental *intellectual* problem of knowledge and understanding: How does the God-of-Value in fact fit into the God-of-Power? How does our human world in fact fit into the physical universe? And it is, third, our fundamental *philosophical* or *conceptual* problem: How is it possible for the God-of-Value to exist in the God-of-Cosmic-Power? How can consciousness, free will, our experienced world, everything of value possibly exist embedded in the physical universe?

All our other problems – philosophical, intellectual, scientific, artistic, technological, personal, social, political, economic, educational and moral – are, in one way or another, subordinate or specialized parts of these three fundamental, religious problems. This is fundamentalism with a vengeance!

Unfortunately, our dishonestly semi-religious and secular world does not see things in this way. Having failed to cut God in half, cleanly and bravely, our world fails to appreciate that the task of trying to put the two halves together again *is* our fundamental problem, underpinning all our other problems, everything else that we do. As we shall see, this failure has damaging consequences for science, for intellectual inquiry more generally, for education, for art, for politics, for the richness of our lives, for everything we strive for and do, for our whole human world, for life on the planet. Because we fail to grasp our fundamental *need* to help the God-of-Cosmic-Value to flourish inside the God-of-Cosmic-Power, not surprisingly we fail to realize – to apprehend and create – *in reality*, what is of most value, actually and potentially, in the circumstances of our lives.

In this chapter I indicate how our failure to appreciate that our fundamental problem is to help the God-of-Value to flourish within the God-of-Cosmic-Power has resulted in a kind of academic inquiry that is damagingly irrational when judged from the standpoint of the best interests of humanity. Subsequent chapters will fill in details, and will spell out other, related damaging consequences, for natural science, for social inquiry, for politics, for our ideas about what is of value, for ideas about rationality, for education, for philosophy, for our understanding of how our human world fits in to the physical universe.

In the next chapter I show how our experiential world (the God-of-Cosmic-Value) can exist embedded in the physical universe (the God-of-Cosmic-Power). In chapter four I put forward a conjecture about the nature of the God-of-Cosmic-Value – the nature of what is ultimately of value in existence – and discuss problems that this conjecture faces. Chapter five explores questions about the nature and existence of the God-of-Cosmic-Power. I argue that a proper understanding of science reveals that science has already established that the God-of-Cosmic-Power exists in so far, that is, that science is able to establish anything theoretical at all.

In chapter six I argue for "wisdom-inquiry" – a kind of academic inquiry rationally designed to help us realize what is genuinely of value in life embedded as we are in the physical universe (a kind of inquiry designed to help the God-of-Cosmic-Value to flourish within the God-of-Cosmic-Power, in other words).[1] This develops further, and supplements, the account of, and arguments for,

---

[1] What is wisdom? There can be no such thing as *the correct* definition of "wisdom", as Popper in effect established long ago: see Popper (1962), vol. 2, Chapter 11, section ii. "Wisdom" may quite legitimately mean a variety of things, depending on context, and the aim we have in mind. Here – as I indicate in chapter 6 – by wisdom I mean the capacity, and active desire, to realize what is of value in life, for oneself and others, wisdom thus including knowledge, technological know-how and understanding, but much else besides. Given this notion of wisdom, we can say that the basic intellectual aim of inquiry, according to wisdom-inquiry, is not knowledge merely, but rather wisdom. For a slightly more elaborate characterization of wisdom along these lines see my (1984), p.66, or (2007a), p. 79.

wisdom-inquiry I give below, in the present chapter. In chapter seven I tackle the key problem of how *we* can exist and act, with some degree of free will, granted we are an integral part of the physical universe, and thus subject, in all that we think and do, to iron physical law – the all-embracing diktats of the God-of-Cosmic-Power. In chapter eight I indicate how Darwin's theory of evolution needs to be reinterpreted so that it helps explain how life of value has gradually emerged in the physical universe – Darwinian evolution, thus understood, seamlessly merging into history. In chapter nine I indicate how wisdom-inquiry would help us resolve our global problems in wiser, more cooperatively rational ways.

Elementary Moral Implications

What ought you to do in your life in order best to help the God-of-Value to flourish? No one can know you as intimately as you know yourself. No one is able to animate you, guide you, divine your best interests, your deepest longings, as perceptively and effectively as you yourself are. Therefore, your prime religious duty, in helping the God-of-Value to flourish inside the God-of-Cosmic-Power must be to devote yourself to the flourishing of that bit of the God-of-Value that is *you* – that is what is best in you. Your prime religious duty must be to devote yourself to the flourishing of your own best interests. Shockingly, when viewed from the perspective of traditional religions, selfishness not selflessness is our first religious duty.

But this stark primary religious duty of selfishness needs to be qualified in various ways.

First, not everything we desire is desirable. To put it at its most extreme, the drug addict may fiercely desire a fix, but this may not be desirable. Our primary obligation must be to try to discover and pursue, not just what we desire, but that which is most desirable in the circumstances of our life.

Second, we will fail miserably to realize what is genuinely desirable if we ignore what others desire. Without friendship and love, life would scarcely be worth living. Much that is of value in life arises when we collaborate with others. Almost everything of value in our lives we owe to others. Our very identity, even our

18

consciousness, we owe to others – to those who brought us up, taught us to speak and, more distantly, to those, long ago, who created language and human culture. Psychopaths are alone in this world – but even psychopaths who achieve social success, discover how to play by the rules and conceal their cold and absolute indifference to the interests of others. Mere selfishness defeats self interest.

Third, even enlightened self-interest is not enough. Others are of intrinsic value, just as I am, or you are. Another person is another me, another you. A stranger may mean nothing to me personally, and may in no way enrich my life, and yet the interests and concerns of that other person deserve my consideration. I ought not to ride roughshod over the interests of others even when I gain nothing for myself thereby. We may not all be of equal value: is Hitler or Stalin equal in value to Mozart, Gandhi or Einstein? But it is vital that we are all equal before the law. And if we genuinely value ourselves, the chances are that we will neither belittle nor inflate our own value in the general scheme of things.

And fourth, it is of course, not always true that we are best able to care for ourselves. This is not true of babies and young children, the ill, the severely handicapped, the mad, the very old. These must be cared for by others. In fact, whoever we are, circumstances are almost bound to arise in which others know where our best interests lie better than we do ourselves.

Having the interests of the God-of-Cosmic-Value to heart demands that we have our own best interests to heart, but also the interests of others, those we love, our friends and acquaintances, those who depend on us, and even strangers in so far as we are able, the ideal being to realize what is of value freely and cooperatively with others.

The poles of morality are not selflessness and selfishness, as some traditional religions teach, but rather free cooperation with others, and its opposite, the subversion and violation of free, equal cooperation.

Irrational Institutions of Learning
Our fundamental religious task in life, then, is to help the God-of-Value to flourish within the God-of-Cosmic-Power, this task

radiating out from concern for ourselves and those close to us. At present, however, our ability to perform this fundamental life task is adversely affected by our general failure to appreciate the need to cut God in half, and think through the implications of that act. In order to achieve what is of value in life we need to *learn* how to do it. And for that, in turn, we need schools and universities devoted to the task. It is this that we do not at present have. Judged from this perspective, academic inquiry, as I have already indicated, is damagingly irrational. As we shall see, *three* of the four most elementary rules of reason conceivable are violated. And this is a direct consequence of the historical failure to cut God cleanly in half and then work out how to reassemble the pieces as best we can.

This irrationality of our institutions of learning is no mere formal matter. It has had devastating consequences for humanity. It is the underlying cause of a great swathe of unnecessary human suffering and death. For it has prevented us from developing what we so urgently need, institutions of learning rationally devoted to helping us nourish and promote what is of value, actually and potentially, in our lives. In order to avoid the horrors experienced by so many throughout the 20th century – the unimaginable horrors of modern warfare, millions dying as a result of war, death camps, mass starvation, oppression, exploitation and enslavement – we would have needed to learn how to do it. We would have needed to learn how to resolve our problems and conflicts in more cooperatively rational ways than we have in the past, and than we still do at present. And for that, we would have needed traditions and institutions of learning, schools and universities, well-designed and rationally devoted to promoting such learning. It is this that we have not had, and still do not have. The irrationality of inquiry has prevented its development. And this in turn stems from our long-standing failure to cut God neatly in half, grasp that our fundamental problem is to learn how to help that which is of value to flourish in the physical universe, and create institutions of learning well-designed to achieve that end.

We have, it is true, created a kind of inquiry extraordinarily successful in improving our knowledge and understanding of the universe – namely modern science. But this has, in some respects, served to intensify the danger, the suffering and death. The immense success of modern scientific and technological research has vastly increased our power to act. It has led to modern industry, agriculture and medicine, and to a multitude of products of great human value. Those of us who live in the wealthy, industrially advanced nations lead lives incomparably richer and freer than those who lived a couple of centuries ago thanks, in large measure, to the success of modern science and technology. But the vast increase in the power to act, engendered by science and technology, makes possible action that causes harm, suffering and death just as readily as action designed to do good. In the absence of institutions of learning designed to help us acquire wisdom, it is inevitable that we will fail to learn how to use our unprecedented new powers wisely and well. Just this is what the record of the 20th century reveals. The new, terrible technology of war – whether conventional, chemical, biological or nuclear – leads to the death of millions. Industrialization leads to global warming. Modern agriculture, allied to population growth, leads to destruction of tropical rain forests and other natural habitats, and to the mass extinction of species. Vast discrepancies in wealth develop across the globe because some regions have long benefited from modern science and technology, while others have not. At the same time, traditional ways of life, languages and cultures are swept aside and crushed as an incidental consequence of the remorseless spread of the modern world. It is hardly too much to say that *all* our current global problems are the result of the exercise of our new, unprecedented powers, made possible by modern science and technology, in the absence of a more fundamental concern to learn how to use these powers wisely and well, for the benefit of all people, now and in the future.

The extraordinary success of modern science and technology puts humanity into a position of unprecedented peril. Professor Martin Rees, President of the Royal Society, even thinks that this may be "our final century" (the title of a book of his[2]). There is hardly any more urgent task, as far as the long-term welfare of humanity is concerned, than to develop traditions and institutions

---

[2] Rees (2003).

of learning, schools and universities, rationally devoted to helping us tackle our conflicts and problems of living in wiser, more cooperative ways.

One can put the point quite simply like this. Our only hope of tackling our immense global problems in humane and successful ways is to tackle them *democratically*. But this in turn requires that electorates of democracies have a good understanding of what our global problems are, and what needs to be done to resolve them humanely and successfully. We cannot expect democratically elected political leaders to do – or to be able to do – what needs to be done to resolve (or help resolve) global problems if those who elected them are, by and large, ignorant about what the problems are, and what needs to be done in response to them. Widespread understanding about our problems is a prerequisite for tackling these problems successfully. But this in turn requires that democracies possess substantial programmes of public, adult education about what our problems are, and what we need to do in response to them. This cannot come from politicians, the media, the priesthood, or private individuals (although all these can play a part). It must come from universities. Universities need to take up, as a primary task, public education about what needs to be done in response to our immense problems of living. And "public education" means, not "instruction from on high", but rather debate, argument, learning and teaching in both directions, good ideas for action being seized upon and publicized wherever they may come from, universities stirring up informed and intelligent debate in the public arena. But this is hardly what universities set out to do today. Their primary task, rather, is to accumulate expert, specialized knowledge – thus making, in some respects, our situation even worse. Judged from the highly traditional standpoint of helping to promote human welfare, universities today fail to do what they most need to do. They are, as we shall see in detail below, damagingly and dangerously irrational. Universities fail dreadfully to help humanity learn how to enable the God-of-Cosmic-Value to flourish within the God-of-Cosmic-Power.[3]

---

[3] There are some recent signs, however, that changes are underway in universities, the task of tackling problems of living being given greater

Remarks concerning Intellectual History

How did this disaster come about? How did our institutions of learning in fact develop, and how ought they to have developed?

Once upon a time, in medieval Europe, science and scholarly inquiry, in so far as it existed, in schools, monasteries and rare universities, had an overall Christian ethos and purpose. Christian theology reigned supreme, and other branches of inquiry had to find a place within that overall framework. A basic task for philosophy was to discover how to reconcile the thought of the ancients with Christian thought, the emphasis being first on Plato, then Aristotle.

Free thought was severely curtailed by religious authorities. Cutting God in half was simply not an option. Men were put to death for expressing far less challenging ideas. Giordano Bruno was burned at the stake in 1600 for suggesting that the universe is infinite.[4] About the same time, Domenico Scandella, known as Menocchio, an obscure peasant, a miller, from northern Italy, was put to death for heresy. Menocchio held that once "all was chaos, that is, earth, air, water and fire were mixed together; and out of that bulk a mass formed – just as cheese is made of milk – and worms appeared in it, and these were the angels".[5] Galileo was forced to recant his support for the Copernican view that the earth goes round the sun (and not the sun round the earth).[6] Scholarly inquiry was firmly in the grip of the Church, and those who deviated from Christian doctrine were punished.

---

emphasis, as we shall see in chapter 9.

[4] For accounts of Giordano Bruno's thought, life and death see: White (2006); Gatti (1999). There is some doubt as to whether Bruno was burnt at the stake for his theological or cosmological views, but it would seem that the latter did have a role to play.

[5] For a wonderful account of Menocchio's beliefs, life and death see Ginzburg (1980). See pp. 5-6 for a more detailed statement of Menocchio's beliefs, from which the quotation in the text is taken.

[6] See Shea and Arigas (2003); McMullin (2005); White (2007). Not everyone agrees that Galileo was accused of heresy because of his advocacy of Copernicanism. Pietro Redondi claims that Galileo's heresy was his espousal of atomism which contradicted the doctrine of the Eucharist: see Redoni (1988).

Those responsible for the birth of modern science in the 16th and 17th centuries – Copernicus, Galileo, Kepler, Descartes, Huygens, Boyle, Hooke, Newton, Leibniz and others – had to exercise great care not to offend secular and religious authorities. Galileo argued that the Bible and the Church had authority to pronounce on the moral and spiritual realm, but not on the realm of Nature, which natural philosophy should be free to study, unencumbered by outside interference. But Galileo failed dramatically to convince the ecclesiastical authorities of his case, and subsequently natural philosophers had to proceed diplomatically, and with caution, in order to avoid suffering Galileo's fate – let alone the far worse fate of Bruno.

It would be wrong, however, to think that modern science was born in opposition to Christianity. All those who played a major role in the creation of modern science were sincere Christians, Catholic or Protestant. The Catholic Church itself supported scientific research. Jesuits in Rome, in particular, engaged in scientific enquiry. Major contributors to the birth of modern science, from Galileo to Descartes and Newton, thought long and hard about the relationship between the new science, and Christianity. Christianity even provided something like a rationale, a justification, for the new science. First, there was the idea that God, being good, and having the best interests of humanity to heart, would have ensured that the natural world is comprehensible to man, so that he can, by his own efforts, improve knowledge and understanding of nature. Christianity promised that science – or natural philosophy as it was then called – could meet with success. Second, there was the idea that pursuing natural philosophy could be justified on the basis of the Christian virtue of charity: new knowledge would result in new medicines, new technology, which would relieve human suffering and be of great benefit to humanity in other ways. Modern science was born and grew up within the framework of Christianity.[7] The break with religion, in so far as it has happened, came later.

Gradually, modern science established its independence from religion, along the lines anticipated by Galileo. Laplace, around

---

[7] See Hooykaas (1977).

1800, when taken to task by Napoleon for not mentioning God in his great Newtonian treatise *Mécaniqu Céleste*, replied that he had no need of that hypothesis.[8] In the 19th century it had become possible for T. H. Huxley, Darwin's bulldog, openly to express agnosticism (a term Huxley invented) – although Darwin's theory of evolution, with its suggestion that humanity is descended from apes, provoked an outcry from religious quarters. But neither Darwin nor Huxley were burned at the stake.

A major step towards the development of a secular academic inquiry, free from religious influence or interference, happened in the 18th century with the rise of the Enlightenment. It was during the Enlightenment that the idea began to spread that all ideas should be open to criticism, and not just ideas about the physical universe. Nothing should be decided merely on the basis of authority and tradition, whether religious or secular: everything should be open to assessment by means of experience and reason. The *philosophes* of the French Enlightenment in particular – Voltaire, Diderot, Condorcet and others – had the profound and magnificent idea that humanity should seek to learn from scientific progress how to achieve social progress towards an enlightened world.[9] As we shall see, implemented properly, this would have resulted in a kind of inquiry rationally designed to help humanity learn how to achieve what is of most value in life (thus helping the God-of-Value to flourish within the God-of-Cosmic-Power). But the *philosophes*, despite having their hearts in the right place, botched the job of putting their basic idea properly into practice. Impressed by the progressive success of natural science, they thought, mistakenly, that the task was to develop the social sciences alongside the natural sciences. This idea was developed further throughout the 19th century, and put into institutional practice in the first half of the 20th century with the creation of departments of social sciences in universities throughout the developed world – departments of economics, anthropology, sociology, psychology, political science, international affairs,

---

[8] See Dampier (1971), p. 181.
[9] The best account known to me of the Enlightenment – especially the French Enlightenment – is Gay (1973).

linguistics. Theology shrank from being the dominant branch of inquiry to becoming a kind of intellectual fossil, tolerated as a token gesture to the past, in occasional university departments. (It has, today, a more prominent role, perhaps, in the USA than it does in Europe.) By the early 20th century, academic inquiry had become wholly secular in character, free of Christian influence.

The outcome of all this is what, by and large, we have today: a kind of inquiry that may be called *knowledge-inquiry*. The fundamental humanitarian aim of knowledge-inquiry is to help promote human welfare. Academic inquiry does this by pursuing the *intellectual* goals of knowledge, understanding and technological know-how. *First*, knowledge is to be acquired; once acquired, it can be applied to help solve social problems. Thus, throughout academic inquiry there runs the distinction between "pure" and "applied" inquiry. (There is even "applied" philosophy.) The primary task of academic inquiry is to solve problems of knowledge; once knowledge has been acquired, this can be applied to help solve social problems, problems of living.[10]

----

[10] How, then, did the failure to cut God in half distort the development of academic inquiry? I tackle this question towards the end of this chapter. In summary, my answer goes like this. Academic inquiry in Europe began within the framework of Christianity, informed and constrained by Christian doctrine. When modern science began, in the 16th and 17th centuries, it was difficult enough for natural philosophers to establish the right to explore factual question concerning the nature of the universe freely, unconstrained by the Church. It would have been quite impossible to advocate cutting God in half – quite impossible to explore openly questions about what is of value in life and how it should be achieved in a way which challenged Christian doctrine, ecclesiastical and secular authority. Only in the 18th century, with the Enlightenment, did this begin to be possible. But by then, secular "rationality" had become identified with science and the pursuit of knowledge. The *philosophes* of the French Enlightenment took their task to be to create social science alongside natural science. Thus was academic inquiry developed as the pursuit of *knowledge*. Thus did the long-standing failure to cut God in half lead to a mode of inquiry which failed to give intellectual priority to the task of helping humanity learn to live wisely – so as to realize what is genuinely of value in life.

26

Knowledge-inquiry is, however, severely and damagingly irrational when judged from the standpoint of helping to promote human welfare. As I have already indicated it violates three of the four most elementary rules of rational problem-solving one can think of. Modify knowledge-inquiry just sufficiently to ensure that all *four* rules are built into the intellectual/institutional structure of universities, and a dramatically different kind of inquiry emerges, which I shall call *wisdom-inquiry*. During the first decade of the 21st century, some changes have taken place in universities which have nudged things a bit towards wisdom-inquiry, largely in response to the perceived urgency of environmental problems, as we shall see in chapter nine. These changes have, however, been *ad hoc*, piecemeal, marginal, and have been undermined by other developments which have tied academia even closer to government, industry and commerce. Overwhelmingly, knowledge-inquiry still dominates academia.[11] It is still widely thought to constitute *rational* inquiry (despite being nothing of the kind). At the time of writing, there is no general awareness of the magnitude and urgency of the changes that are required. If ever the transition from knowledge-inquiry to wisdom-inquiry occurs, this will be an intellectual revolution comparable in significance and impact to the Renaissance, the scientific revolution of the 17th century, or the Enlightenment of the 18th century.

Knowledge-inquiry imposes severe constraints, a kind of harsh censorship system, on academic inquiry.[12] Only that which is

---

[11] See Maxwell (1984) ch. 6; or better, Maxwell (2007a), ch. 6. In the second edition of this book, published in 2007, I look at six aspects of academic inquiry, and compare what I found in 1983 with what I found twenty years later, in 2003. Even in 2003, knowledge-inquiry is dominant in academic practice, although some changes have nudged things a bit, perhaps, towards wisdom-inquiry.

[12] An internal system of intellectual censorship takes over from external ecclesiastical and secular censorship. I should add, however, that even a genuinely rational conception of inquiry, put into practice, would amount to a system of censorship (in excluding irrational "bad" potential contributions). We have here an indication of why it is so important to ensure that intellectual standards that govern academic life are genuinely rational, and of the best. If this is not the case, good potential

relevant to the pursuit of knowledge can be permitted to enter the intellectual domain of inquiry: claims to knowledge, theories, reports of observations and experiments, logic, mathematics, arguments designed to establish or to criticize claims to knowledge. Feelings, desires, values, works of art, political programmes, philosophies of life, proposals for action, views about what our problems of living are and how we should go about solving them: all such things as these, not being directly relevant to the acquisition of knowledge must, according to knowledge-inquiry, be ruthlessly excluded from the intellectual domain of inquiry – although of course factual knowledge about such things can be included. All these personal, social, political and evaluative factors must be excluded from inquiry so that authentic, objective factual knowledge is acquired, and the pursuit of knowledge does not degenerate into the promulgation of mere propaganda and ideology. In order to serve the interests of humanity, inquiry must strive to procure authentic factual knowledge, and must forego the task tackling problems of living directly – the tasks of articulating problems of living and proposing and critically assessing possible solutions, possible human actions – as this involves invoking all those personal, social and evaluative issues which would subvert the acquisition of knowledge. Paradoxically, in order to help humanity solve its problems of living in the best possible way, inquiry must forego trying to solve these problems directly, and must concentrate instead on trying to solve *problems of knowledge*. (Once knowledge is acquired it can of course be made available to help solve social problems.)

Science, according to knowledge-inquiry, employs an even more severe system of censorship. Only *empirically testable* claims to knowledge can be allowed to enter the intellectual domain of science – scientific journals, texts, lecture courses, seminars. Not only must expressions of feelings, desires and values be excluded from science, according to knowledge-inquiry: *metaphysical* propositions must be excluded as well – propositions like "all

---

contributions will be excluded, and potentially good research will be discouraged or blocked.

events have a cause" which can neither be verified nor falsified by means of observation or experiment. [13]

The Enlightenment was not alone in influencing the way academic inquiry developed. The Enlightenment provoked Romanticism, as a kind of backlash, that great movement of ideas and feeling inspired by such figures as Blake, Beethoven, Schubert, the young Goethe, Schiller, Rousseau, Wordsworth, Keats, Hazlitt, Coleridge. Whereas the Enlightenment valued science, knowledge, fact, reason, method, logic, experience, Romanticism found all this oppressive, and prized instead art, imagination, passion, inspiration, originality, genius, spontaneity, self-realization. This too had an impact on some parts of academic inquiry, primarily in the humanities: literary and culture studies, feminist studies, philosophy, history, psychology and sociology. It led to various anti-scientific and anti-rationalist movements, such as existentialism, post-structuralism and post-modernism, and to academic work dedicated to political causes, whether of left or right. [14]

Academic inquiry today is thus a sort of mixture of Enlightenment and Romantic influences, more or less at odds with each other. But Romanticism has exercised only a marginal influence. Knowledge-inquiry is still the dominant conception of inquiry, and, as I have mentioned, the only influential idea as to what constitutes *rational* inquiry. [15] Schools of thought influenced by Romanticism tend to be anti-rationalist.

Damaging Irrationality of Knowledge-Inquiry

But knowledge-inquiry, when judged from the standpoint of helping us to realize what is genuinely of value in life is, as I have said, so grossly irrational that it violates three of the four most basic rules of reason conceivable.

---

[13] For a much more detailed exposition of knowledge-inquiry (or "the philosophy of knowledge" as I have called it) and standard empiricism, see my (1984), ch. 2, second edition (2007a).

[14] See Berlin (1979), ch. 1; Berlin (1999).

[15] For grounds for holding knowledge-inquiry is dominant in universities, see work referred to in note 11.

What do I mean by "reason"? As I use the term, reason appeals to the idea that there is some no doubt rather ill-defined set of rules, methods or strategies which, if put into practice, give us our best chance of solving our problems, realizing our aims, other things being equal. The methods of reason do not guarantee success; nor do they determine what we should do. They help *us* to discover and decide what to do, and do not decide for us.

Four absolutely basic rules of reason are the following:-

(1) Articulate, and try to improve the articulation of, the problem to be solved.

(2) Propose and critically assess possible solutions.[16]

(3) When necessary, break recalcitrant problems into easier-to-solve preliminary, subordinate, specialized problems.

(4) Interconnect basic and specialized problem-solving so that each may guide the other.[17]

Any problem-solving enterprise which persistently violates any one of these basic rules must be seriously irrational, and must be damaged as a result. Knowledge-inquiry – and academic inquiry as it exists today – violates *three* of these four most basic rules of reason.

---

[16] Karl Popper devoted much of his working life to establishing the fundamental importance of the first two of these four rules of rational problem-solving. His philosophy of science is a particular application of these two rules to the special case of science. In one place he formulates them like this: "the one method of all *rational discussion*, and therefore of the natural sciences as well as philosophy ... is that of stating one's problem clearly and of examining its various proposed solutions *critically*": Popper (1959), p. 16. Popper was, however, too vehemently opposed to specialization to appreciate that it is (when problems are intractable) a vital component of rationality – it being possible to counteract potentially harmful effects of rule (3) by implementing rule (4) as well. Popper's *critical rationalism* thus consists of (1) and (2), but omits rules (3) and (4). For Popper's anti-specialism see his (1963), p. 136, or his remark that "If the many, the specialists, gain the day, it will be the end of science as we know it – of great science. It will be a spiritual catastrophe comparable in its consequences to nuclear armament" Popper (1994), p. 72.

[17] The fundamental importance of interconnecting specialized and fundamental problem-solving is argued for in some detail in my (1980).

Granted that the basic aim of inquiry is to help promote human welfare, help humanity achieve what is of value, then the basic problems that need to be solved are problems of *living*, problems of *action*, not problems of *knowledge*. It is always what we *do*, or refrain from doing, that enables us to achieve what is of value, and not new knowledge (except when knowledge is sought for itself, as being itself of value). Even when new knowledge and technology are needed to achieve what is of value – as they are in connection with medicine, for example – it is always what the knowledge and technology enables us to *do* that achieves what is of value – health – and not the knowledge and technology in themselves. Thus, if academic inquiry is to accord with the most elementary requirements of reason conceivable, it must give absolute intellectual priority to the two tasks of:-

(1) Articulating, and trying to improve the articulation of, our problems of living, our problems of action;

(2) Proposing and critically assessing possible solutions – possible actions, policies, strategies, plans, political programmes, legislation, institutions, religious views, philosophies of life.

But it is just this which knowledge-inquiry, as we have seen, cannot do. Knowledge-inquiry gives intellectual priority to tackling problems of *knowledge*, not problems of *living*. Strictly speaking, it is even worse than this: knowledge-inquiry is actually *restricted* to tackling problems of knowledge: as we have seen, it is *prohibited* from articulating and trying to solve problems of living. Knowledge-inquiry violates the two most elementary rules of reason imaginable.[18]

---

[18] Things are, at the time of writing, not quite as bad in universities as this might suggest. Once upon a time, in the 1980s for example, knowledge-inquiry was much more strictly adhered to. Increasingly, during the last decade or so, universities have created departments, institutes and centres devoted to policy studies, devoted to tackling major social and global problems, research being organized in response to policy issues, social and environmental problems. Increasingly, in other words, universities have moved away from implementing knowledge-inquiry in a strict, rigorous, thoroughgoing fashion, as I have already indicated, and as I shall discuss at greater length in chapter 9. Nevertheless, knowledge-inquiry still exercises a pervasive influence

Knowledge-inquiry, as pursued in universities today, does, however, put rule (3) into practice to a quite extraordinary extent. The outcome is the maze of specialized disciplines that is such a striking feature of academic inquiry – each discipline containing a multitude of sub-disciplines, each in turn containing a multitude of sub-sub-disciplines, and so on.

But, having failed to implement rules (1) and (2), knowledge-inquiry cannot implement rule (4). It cannot interconnect tackling of specialized problems of knowledge and technology with tackling of basic problems of living because consideration of the latter is excluded from the intellectual domain of inquiry.

In short, when judged from the standpoint of helping to promote human welfare – of helping us realize what is of value in life – by intellectual and educational means, knowledge-inquiry is so profoundly irrational that it violates *three* of the four most elementary rules of reason one can think of. And this violation of reason is no minor matter: it is wholesale and structural, and has immense, long-term damaging consequences.

The failure to implement rule (4) means that specialized research develops in ways unrelated to, and not influenced by, active tackling of more fundamental problems of living, which means, of course, that priorities of scientific research will tend to come to reflect the interests of scientists themselves, and of governments and industry (who pay for the research), rather than reflecting the priorities of real human need. Thus medical research comes to reflect the interests of drug companies and perhaps the health interests of people living in wealthy countries, and not the health interests of the poor of the world. Vast sums are devoted to

---

over much academic research and teaching. Policy studies, the direct tackling of problems of living, still tend to be pushed to the fringes. Knowledge-inquiry is still the only agreed conception of *rational* inquiry. Its damaging *irrationality* has still not been generally appreciated and understood. There is at present no active campaign to abolish knowledge-inquiry and transform universities so that they come to take, as their primary task, to help us tackle our problems of living in increasingly cooperatively rational ways – above all, our major global problems of climate change, war, poverty, destruction of natural habitats and extinction of species.

military research, as if our war torn planet really needs ever more lethal means for delivering death.[19]

But far worse is the failure to implement rules (1) and (2). This means academia fails to give intellectual priority to the tasks of articulating our conflicts and problems of living and proposing and critically assessing increasingly cooperative, just and peaceful resolutions - resolutions which, if enacted, lead to the realization of what is of value in life. Instead the resources of research are devoted to increasing knowledge and technological know-how which, as I have already said, increases our power to act, thus leading to much good but also much harm. *In the absence of the sustained endeavour to discover how we can resolve our conflicts and problems of living in more cooperatively rational, just and peaceful ways than we have done in the past, it is inevitable that the vast increase in the power to act, acquired by some from modern science and technology, will result in such things as rapid population growth, ever more lethal war and the means to engage in such war, global warming, vast inequalities of wealth and power across the globe, destruction of natural habitats and rapid extinction of species, depletion of finite natural resources, destruction of traditional ways of life, cultures and languages, and even epidemics such as that of aids* (aids being spread by modern methods of travel). These modern crises, so characteristic of our age, are all the product of giving priority, in the long term, to tackling problems of *knowledge* over problems of *living*. They are the product of the long-standing, almost universally overlooked wholesale *irrationality* of academic inquiry – of our most intelligent and influential public thought.[20]

The key to creating a better world is to create better thinking about how to do it. We need to bring about a revolution in academic inquiry so that it devotes elementary rules of reason to the task of helping humanity solve its problems in more cooperative, just and peaceful ways. Thinking intelligently about

---

[19] See Smith (2003); Langley (2005).

[20] For a more detailed demonstration of the damaging irrationality of knowledge-inquiry when judged from the standpoint of helping to promote human welfare, see my (1984) or (2007a), ch. 3.

how to solve our problems is clearly not *sufficient* to solve them, but it is, I maintain, *necessary*. We cannot hope to do it employing grossly irrational, malformed modes of thought.

Wisdom-Inquiry

But if knowledge-inquiry is ill-equipped to help us solve our problems of living, realize what is of value in life, what kind of inquiry would be well-equipped to do this? The answer: knowledge-inquiry modified just sufficiently to ensure that all four of the above rules of reason are put into practice. I shall call inquiry that does this *wisdom-inquiry*. It is depicted in diagrams 1 and 2. Its main features are as follows.

First, as a matter of absolute intellectual priority, wisdom-inquiry puts the first two rules of reason into practice. The fundamental intellectual activity is to:

(1) Articulate, and seek to improve the articulation of, those personal, social and global conflicts and problems of living we need to resolve in order to realize what is of value in life;

(2) Propose and critically assess possible increasingly cooperative *actions* designed, if performed, to enable us to solve our problems, realize what is of value in life.

The fundamental intellectual task of wisdom-inquiry, in other words, is to explore, imaginatively and critically, what we might do in order to help the God-of-Cosmic-Value to flourish within the God-of-Cosmic-Power. It is to devote reason to the task of solving our fundamental religious problem in life. This task, at the heart of the academic enterprise, is undertaken by social inquiry and the humanities.

Next, wisdom-inquiry puts rule (3) into practice. Two profoundly recalcitrant problems that confront us are (a) the scientific problem of acquiring knowledge and understanding of the universe, and ourselves as a part of the universe, and (b) the socio-economic-political problem of creating a better world. In seeking to solve, or help solve, these two basic, recalcitrant problems, wisdom-inquiry:

(3) Tackles a host of subordinate, specialized problems – thus creating specialized disciplines of the natural, technological and

formal sciences, social inquiry and the humanities (see diagrams 1 and 2).

But, in order to try to ensure that specialized problem-solving is pursued in such a way as to help solve fundamental problems such as (a) and (b), wisdom-inquiry also puts rule (4) into practice as well. That is, wisdom-inquiry:

(4) Interconnects basic and specialized problem-solving so that each may guide the other (indicated by two-headed arrows in the diagrams 1 and 2).

What, then, are the main differences between knowledge-inquiry and wisdom-inquiry?

One big difference has to do with *problems* – what they are, and how they are tackled. Knowledge-inquiry restricts itself, in the first instance, to tackling problems of *knowledge*. The tackling of problems of living is very much a secondary matter, and tends to be restricted to developing and providing relevant knowledge and technological know-how. Furthermore, knowledge-inquiry, as implemented in academic practice in universities, tends to exclude the vital task of articulating, and improving the articulation of, problems of knowledge from the intellectual domain of inquiry (which tends to be restricted to contributions to knowledge). In addition, knowledge-inquiry tends in academic practice to result in the tackling of a maze of specialized problems of knowledge, with little discussion of broad, fundamental problems. There is a tendency to violate rules (1), (2) and (4), in other words, even within the domain of knowledge.

By contrast, wisdom-inquiry emphasizes the intellectually fundamental character of problems of *living*, and stresses the vital need to articulate, and improve the articulation of, problems to be tackled (whether problems of living, of knowledge, or of technological know-how). It stresses, too, the vital importance of tackling broad, fundamental problems in addition to the maze of specialized problems, and the importance of allowing work on these two kinds of problems to influence each other, in accordance with rule (4). At the most fundamental level of all, there is the problem: How can the God-of-Value exist and best flourish in the God-of-Cosmic-Power? How can we best help that which is of value associated with our human world to flourish in the real

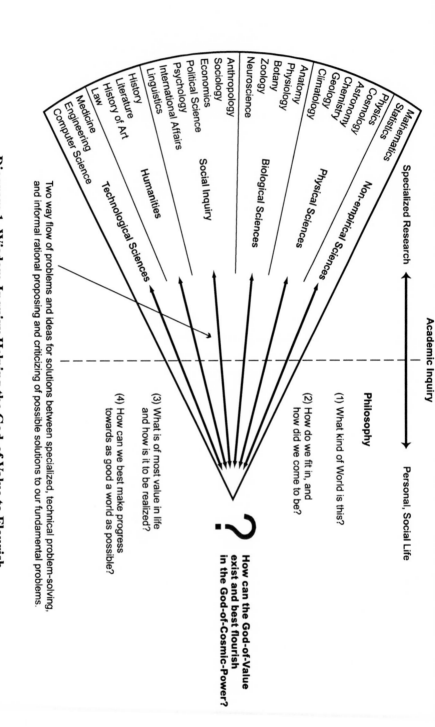

Diagram 1 Wisdom-Inquiry Helping the God-of-Value to Flourish

Academic Inquiry

Specialized Research ↔ Personal, Social Life

Non-empirical Sciences
- Mathematics
- Statistics

Physical Sciences
- Physics
- Cosmology
- Astronomy
- Chemistry
- Geology
- Climatology

Biological Sciences
- Anatomy
- Physiology
- Botany
- Zoology
- Neuroscience

Social Inquiry
- Anthropology
- Sociology
- Economics
- Political Science
- Psychology
- International Affairs
- Linguistics

Humanities
- History
- Literature
- History of Art
- Law

Technological Sciences
- Medicine
- Engineering
- Computer Science

Philosophy

(1) What kind of World is this?

(2) How do we fit in, and how did we come to be?

(3) What is of most value in life and how is it to be realized?

(4) How can we best make progress towards as good a world as possible?

?

How can the God-of-Value exist and best flourish in the God-of-Cosmic-Power?

Two way flow of problems and ideas for solutions between specialized, technical problem-solving, and informal rational proposing and criticizing of possible solutions to our fundamental problems.

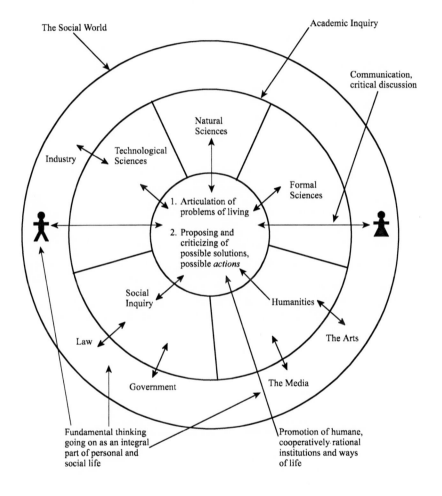

The Social World

Academic Inquiry

Communication,
critical discussion

Natural
Sciences

Technological
Sciences

Industry

Formal
Sciences

1. Articulation of
problems of living

2. Proposing and
criticizing of
possible solutions,
possible *actions*

Social
Inquiry

Humanities

Law

The Arts

Government

The Media

Fundamental thinking
going on as an integral
part of personal and
social life

Promotion of humane,
cooperatively rational
institutions and ways
of life

Diagram 2
Wisdom-Inquiry Implementing Four Rules of
Rational Problem Solving

world? All this needs to be built into the institutional structure of academic inquiry and the university: see diagram 1.

Another big difference has to do with the nature of social inquiry. Given knowledge-inquiry, social inquiry is primarily social *science*. Economics, psychology, sociology, anthropology, political science, linguistics are all pursued primarily as *sciences*, or at least as disciplines concerned to improve knowledge and understanding of aspects of the human world. Concern with policy is subordinate and secondary. Given wisdom-inquiry, social inquiry is not primarily *science*, or the pursuit of *knowledge*. Instead of being concerned to solve problems of *knowledge*, social inquiry has, as its basic task, to help humanity solve those problems of *living* that need to be solved in order that what is of value may be realized. Social inquiry proposes and assesses, not *claims to knowledge*, but rather *proposals for action*. How can democratic world government be formed? What can be done to create sustainable world industry and agriculture? How can poverty be eliminated? These are the kind of problems at the heart of social inquiry. Of course emerging out of, and feeding back into, the fundamental intellectual activity of exploring imaginatively and critically possible personal, social, institutional and global *actions*, there will be a concern to acquire relevant factual knowledge about the human world, in order to assess critically, and in order to discover how to implement, proposals for action. This pursuit of knowledge about social phenomena would, however, be subordinate and secondary to the primary activity of tackling problems of living.[21]

---

[21] It would of course be vital to ensure that knowledge of social phenomena is developed to assess critically proposals for action, and not to provide spurious support for policies already decided upon. Policy may legitimately influence *what* social research is conducted, but cannot influence the *outcome* of such research. And, more generally, values inevitably, and quite properly, influence *what we choose to acquire knowledge about*, but must not be allowed to influence *the content of knowledge*, questions of *fact, truth and falsity*. (Any such influence must go in a negative direction. A new, highly desirable item of technology which will endanger life if not safe and reliable, must be especially severely tested.)

A third big difference between the two modes of inquiry has to do with the status and position of social inquiry, and its relationship to natural science. Within knowledge-inquiry, the different disciplines fall into a natural hierarchy. At the base there is theoretical physics (and perhaps cosmology); then there are various applications of physics and more phenomenological or practical parts of physics, such as astrophysics and solid state physics; then chemistry, and rather more observational physical sciences such as astronomy and geology; then the biological sciences, ranging from molecular biology, physiology, botany, evolutionary theory, to the study of animal behaviour; then come the social sciences, from anthropology to sociology, psychology, economics and political science, with philosophy, perhaps, at the top of the hierarchy. The idea is that any science at a given level, physiology say, may use discoveries at a lower level, such as molecular biology, chemistry, and even, perhaps, physics, but will not use discoveries made at a higher level, such as anthropology or psychology. Academic status tends to be associated with this hierarchy. The nearer the base a discipline is so the "harder", the more scientific, exacting and rigorous, the discipline tends to be judged to be, whereas the nearer the top a discipline is so the "softer", the more unscientific, woolly, and unrigorous it tends to be judged to be. Whereas physics is intellectually fundamental, hard, and fiercely scientific, a social science such as psychology or sociology is not remotely fundamental, and is soft and of questionable scientific status.

Wisdom-inquiry transforms all this entirely. Within wisdom-inquiry, it is social inquiry, as we have seen, that is intellectually fundamental – in that it tackles our fundamental problems of living, of action. All of the natural and technological sciences – including physics – are intellectually subordinate in that they tackle subordinate problems of knowledge, explanation, understanding and know-how. (It is just this that is depicted in diagram 2.) The hierarchy of knowledge-inquiry is turned upside down.

A fourth big difference between the two modes of inquiry has to do with the relationship that each demands should exist between academic inquiry as a whole and the rest of the social world.

According to knowledge-inquiry, the proper, basic *intellectual* relationship is that of academia studying the social world and acquiring knowledge of it. In order to do this properly, academia must take care that the act of studying does not affect what is studied – or the knowledge acquired will be out of date. In addition, as I have already mentioned, academic inquiry must shield itself from the corrupting influence of the social world so that genuine knowledge of factual truth may be acquired. The influences of public opinion, politics, religion, emotion, values, must all be kept at bay, so that the search for knowledge does not degenerate into the promulgation of mere propaganda or ideology. Those who contribute to inquiry must be properly qualified insiders possessing at least a Ph.D. Apart from financial support and appreciation, academia requires primarily its independence from the rest of the social world. Everything is dramatically different given wisdom-inquiry. The basic task of academia becomes to help humanity tackle problems of living in increasingly cooperatively rational ways so that what is of value in life may be realized. Far from studying the rest of the human world and, apart from that, remaining somewhat aloof, academia needs to be engaged in lively two-way communication and argument, both teaching and learning, doing everything to promote more cooperatively rational tackling of conflicts and problems of living. Good ideas as to how to solve problems of living need to be published and promoted wherever they come from, whether their authors have professional academic qualifications or not. A major part of the job of academia is to extract ideas, whether being implemented in practice or not, from the rest of the social world, and filter out those that are good from those not so good. The whole of academia can be seen as a form of specialized thought, arising from non-specialized thinking we engage in as we live as a result of the implementation of rule (3). But in order to counteract the dangers of specialization, rule (4) needs to be put into practice as well: the specialized thought of academia needs both to guide, and be guided by, the non-academic thought we engage in as we live. As I have said elsewhere, academia needs to function as a kind of people's civil service, doing openly for the public what actual civil services are supposed to do in secret for governments.

Academia needs just enough power to preserve its independence from government, industry, the media, public opinion, but no more. Its task is to learn from, argue with and teach the rest of the world, but not to order, to enforce obedience.

A fifth major difference has to do with the *content* of inquiry. Given knowledge-inquiry, the content of academic journals, books, lectures and seminars all has to do with the pursuit of knowledge: claims to knowledge, reports of observations, experimental results and historical documents, proofs, arguments and criticisms designed to establish the truth. Emotion, values, politics, religion, art are all excluded as subverting, rather than contributing to, knowledge - although claims to factual knowledge about these things are of course included within the intellectual domain of inquiry. Given wisdom-inquiry, it is all the other way round. Emotion, values, politics, religion and art all need to be included within the intellectual domain of inquiry. The basic intellectual aim is to help us realize - apprehend or make real - what is of value. Emotion needs to be included, since we can only make value discoveries of our own, as opposed to parrot the value discoveries of others, if we do attend to our own emotional responses to things. Of course not everything that feels good is good; emotion needs to be subjected to critical scrutiny if it is to indicate what is of value. Emotion is an essential but fallible ingredient of what is needed to discover what is of value. Again, values need to be included; we can hardly discover what is of value if all consideration of values is forbidden. But value-claims need critical scrutiny, just as knowledge-claims do. Yet again, politics needs to be included - or at least political issues, policies, programmes and philosophies, since these are possible solutions to our problems of living. Religion needs to be included, since religions make claims about what is of most value in existence and how it is to be attained or realized. All religious doctrines need, however, critical scrutiny: traditional religions, as we have seen, include all sorts of absurdities. Wisdom-inquiry is itself, of course, a supremely religious endeavour, in that it seeks to discover what is genuinely of value in existence and how it is to be realized. It differs from traditional religions in subjecting all knowledge and value claims to critical scrutiny, including the test of experience. Art needs to

41

be included, as revelations of value, and as criticism of false values in comedy and satire. Literature, too, has a role to play in enhancing our capacity to understand our fellow human beings, and perhaps ourselves, as a result of identifying imaginatively with the fictional characters of novels, plays and films. Quite generally, whereas the intellectual domain of knowledge-inquiry is restricted to that which is relevant to the acquisition of knowledge, wisdom-inquiry includes, and gives an intellectually fundamental place to, ideas relevant to the discovery of what is of value, and it is this difference which accounts for the radically different contents of the two kinds of inquiry.

Both modes of inquiry appeal to experience, and are *empirical,* when possible; they differ radically, however, as to what "experience" means, and what it is that experience decides or assesses. For knowledge-inquiry, "experience" means observation and experiment as this arises in science, and what is established or assessed by its means are claims to knowledge. For wisdom-inquiry, "experience" is what we acquire as we live, as we strive to achieve our ends, enjoy and suffer. And what is assessed by its means are not, primarily, claims to knowledge but rather proposals for action, plans, policies, political programmes, philosophies of life. The well-known empirical method of science of putting forward theories which are then tested by means of observation and experiment is extended, within wisdom-inquiry, to include proposing possible *actions* which are then assessed in terms of what we enjoy or suffer when these actions are executed, either in actuality or preferably, at least initially, in imagination. (Trying to learn from actions performed in imagination rather than in reality suffers from the disadvantage that what is learned is less reliable, but has the advantage that blunders are less costly - suffering in imagination being more bearable than suffering in reality, and mistakes in imagination being easier to rectify than mistakes in reality.)

Enough has been said, I trust, to establish that knowledge-inquiry and wisdom-inquiry differ dramatically, in wholesale, structural ways. Every branch and aspect of inquiry is affected by these major differences. It is important to appreciate, nevertheless, that these differences all stem from the simple demand that inquiry, in seeking to help us realize what is of value in life, should observe elementary rules of reason. The differences all come from modifying knowledge-inquiry just sufficiently so that the four most elementary rules of reason conceivable are observed in the pursuit of what is of value in life.

Our extraordinarily successful pursuit of knowledge-inquiry (successful in its own terms), and our failure to develop wisdom-inquiry, one or two centuries ago, has had dire consequences, as I have already in effect indicated. It is this that has permitted scientific and technological research to develop unchecked by a more fundamental concern with our problems of living. It is this which has resulted in a kind of academic inquiry which fails to give priority to the task of discovering how we can resolve those conflicts and problems of living we need to resolve in order to realize what is of genuine value in life. It is hardly too much to say that *the* crisis behind all the others – global warming, rapid population growth, destruction of natural habitats and mass extinction of species, the lethal character of modern war and the threat posed by modern armaments, conventional, chemical, biological and nuclear – is our energetic pursuit of knowledge-inquiry and our failure to implement wisdom-inquiry. We have put all our efforts into increasing our power to act, via increasing our knowledge and technological know-how, and have failed to give priority to learning how to live wisely.

What we have failed to do, in short, is to develop institutions and traditions of learning rationally devoted to helping us solve our fundamental problem of how to help the God-of-Cosmic-Value to flourish within the God-of-Cosmic-Power. It is just this which wisdom-inquiry would be; and it is just this that we have failed so far to create.

Reasons for Failure to Develop Wisdom-Inquiry
How did this monumental failure come about? A part of the answer is contained in the brief history already indicated. Suppose long ago, in the 15th or 16th century, a great religious movement had emerged which emphasized the points made in chapter one. In our world of suffering it is quite impossible that there should be an all-powerful and all-knowing God who is also all-loving; the God

of Power must be split off from the God of Value; the former is impersonal, cannot know what it does, and therefore can be forgiven the terrible things that it does; the latter is suffering humanity; we feel and see suffering, we struggle to combat it, but our efforts are only partly successful, and our fundamental religious task in life is to relieve suffering and help what is genuinely of value in life to flourish. Somehow, let us suppose, the Church gave way before the passionate sincerity of the disciples of this movement, and the blazing cogency of their arguments, and throughout Europe churches and religious services were devoted to celebrating the supreme value of human life, and to discovering how what is of value may realised – in this life, and not in a mythical next one. Mighty landowners, the wealthy and powerful, Princes and Kings, like Bishops and the Pope, could no longer take their power, possessions and offices for granted. What is of supreme value is a mystery; it is in us and around us, in what we do, feel and think, in how we are with each other. It is for each one of us to discover, and to live, as best we can, and not for the high and mighty, the Kings and Popes, to decide for us. If such a religious, political and intellectual revolution had occurred, it would have been recognized, long ago, that our fundamental problem is to discover how to help the God-of-Cosmic-Value to flourish within the God-of-Cosmic-Power. Wisdom-inquiry, rationally designed to help us solve this problem as we live, might well have been the outcome.

But no such religious convulsion occurred in the 15th or 16th century. Those who founded modern science, in the 16th and 17th centuries, found it quite difficult and perilous enough to establish the principle that science – or natural philosophy as it was then known as – should be free to explore ideas about nature without hindrance from religious or secular authority. Bruno, Scandella and Galileo, as we have seen, were put to death or imprisoned for doing just that. Those who developed modern science after Galileo – Descartes, Newton, Huygens, Hooke, Boyle and others – took care, by and large, to restrict their inquiry to matters of fact and steered clear of moral, political and, above all, religious issues that would have provoked secular and religious authorities.

Not till the 18th century, with the rise of the Enlightenment, did the idea that *everything* should be open to critical scrutiny gain widespread currency. But by then *critical, rational inquiry* had come to mean *the scientific pursuit of knowledge.* Newton had achieved unprecedented success in predicting and explaining a multitude of terrestrial and astronomical phenomena by means of one theory – his laws of motion and law of gravitation. An assured *method* had it seemed been discovered which, for the first time in history, enabled man to acquire profound new knowledge of nature – the famous empirical method of science. The *Philosophes* of the French Enlightenment in particular – Voltaire, Diderot, D'Almbert and the rest – were immensely impressed by Francis Bacon's idea that acquiring genuine knowledge of nature would make it possible to transform the human condition, and by Newton's achievements in acquiring such knowledge and demonstrating (it seemed) what methods need to be employed to acquire knowledge.[22] The *philosophes* concluded, understandably enough, that if knowledge of *nature* is important if we are to create a better world, knowledge of *the human world*, of *society*, is even more vital. In order to transform the social world, knowledge of the laws of the social world must, surely, be absolutely essential. So they set about creating the social sciences alongside the natural sciences: economics, psychology, anthropology, sociology and the rest.[23] This idea was further pursued throughout the 19th century, and built into the institutional structure of academia in the early 20th century with the creation of departments of social science all over the world.[24] The outcome is what we have today, *knowledge-inquiry*, natural and social science devoted to the pursuit of knowledge, damagingly irrational when judged from the standpoint of helping humanity learn how to realize what is of value in life – learn how to help the God-of-Cosmic-Value flourish within the God-of-Cosmic-Power.

---

[22] For Newton's influence on the French Enlightenment see P. Gay (1973), vol. 2, 'The Science of Freedom', ch. 3.
[23] See Ibid.
[24] For an excellent, brief account of the origins and development of the social sciences along these lines, see Fargaus (1993), Introduction. See also Hayek (1979).

The *philosophes* of the Enlightenment nearly got it right. They had the profoundly important idea that it might be possible to learn from scientific progress how to make social progress towards an enlightened world. This might indeed be said to be the basic idea of the Enlightenment. Not just in science, but in social and political life too, ideas (in particular, ideas for living) should be decided, not by mere tradition and authority, but by critical rationality and experience. Instinctively, in their lives, the *philosophes* sought to put this creed into practice. But in developing the creed, they blundered. Instead of seeking to apply progress-achieving methods – arrived at by appropriately generalizing the progress-achieving methods of science – directly to *social life*, to the *problems* of social life, the *philosophes* applied these methods to *social science*, to the task of acquiring *knowledge* of social life. They applied reason, not to the task of making *social progress* towards an enlightened world, but rather to making *progress in knowledge* about the social world. The outcome was that social inquiry was subsequently developed as social science, and not as the endeavour to get rational methods of problem solving into social life. As a result, the Enlightenment failed to create wisdom-inquiry, and bequeathed knowledge-inquiry to us instead.[25]

Creating a kind of inquiry devoted, in an effective and genuinely rational way, to helping humanity learn how to create a better world, is almost more difficult to do today than it would have been to do in the past. In earlier times one would have been burnt at the stake or thrown into prison – as one still would today, as it happens, in some parts of the world. But in liberal democracies today, it is not death, torture or imprisonment that one will suffer, but rather sheer indifference and disbelief. How could the still, small voice of reason have an impact on the mighty industry of academia today, confident in its fundamental rectitude, a massive, complex institutional reality no more amenable to rational control

---

[25] In chapter 6 I give a more detailed discussion of what the mistakes of the *philosophes* were in developing and applying their basic Enlightenment idea that we should learn from scientific progress how to achieve social progress towards an enlightened world.

than history, or the motion of the earth? A mere *argument*, however valid, seems powerless against such an institutional juggernaut.

This at least has been my personal experience. For over thirty years I have spelled out the argument sketched above, in and out of print, for the urgent need to bring about a revolution in the aims and methods of academia. Critics have praised and savaged my work. But hardly anything has happened (although very recently things have begun to stir, as we shall see below, briefly, and in more detail in chapter 6). [26] It is not just that the argument has had no discernible effect on academic practice. The argument has not entered the public arena. Even specialists in the field know nothing of it. In so far as they have some vague awareness of it, they know it is of no account precisely because it has been generally ignored. Just as nothing succeeds like success, so too nothing fails like failure.

Despite this, the case for creating wisdom-inquiry has never been so urgent and decisive. The combination of population growth, modern agriculture and industry, and modern armaments, all made possible by the immense technical success of science and technological research, has generated unprecedented global crises. Humanity needs to *learn* how to overcome these crises, but the learning that is involved is the kind provided by wisdom-inquiry, not knowledge-inquiry. Wisdom is no longer a luxury; it is a necessity.

Global Warming and Terrorism

An example is global warming. This is caused by the consumption of coal and oil for energy and travel, all made possible by modern science and technology. There are some who, even at the time of writing (2008) doubt that global warming is caused by humanity. But this is hardly the point. It has been known for one and a half centuries that carbon dioxide is a greenhouse gas. For decades, there has been no doubt whatsoever that the amount of CO_2 in the atmosphere is increasing, and an increase in CO_2 will lead to global warming. Even if the sceptics were right – which

---

[26] See also note 18.

47

they are not – and recent increase in global temperatures are due to natural causes, all the more does it make it a vital matter to cut back on CO_2 emissions so as not to intensify global warming.[27]

We have known for decades that industrial development will lead to global warming. Science has made it possible for us to cause global warming, and science has also, of course, made it possible for us to find out that we are the cause. What science cannot do is solve the immense political, economic, industrial and social problems that need to be solved if we human beings are to cut back sufficiently on our global emissions of CO_2. Knowledge-inquiry is simply not designed to help solve these immense problems of living. But wisdom-inquiry is so designed. It puts problems of living at the heart of the academic enterprise. Wisdom-inquiry would be centrally and fundamentally concerned to propose and critically assess ideas about what should be done to cut back on CO_2 emissions. It would energetically hunt for the best ideas for action around, wherever they may be found, and would energetically publicize these ideas in relevant circles of influence – political, industrial, commercial, educational, and in the media. All this would be done in a way which links up with the multitude of scientific and technological issues that need to be tackled to develop new sources of power and less consumption of it, and new modes of transport.[28]

Another example is terrorism. 9/11 is, of course, a terrible illustration of the way in which modern technology can be exploited to cause havoc. Even though the terrorist themselves were armed with nothing more sophisticated than knives, they were able to exploit aeroplanes and modern buildings to kill some three thousand people. In the future, no doubt, terrorists will be able to get their hands on more lethal weapons – biological or

---

[27] For an excellent history of the discovery of global warming see Weart (2003).

[28] For an excellent recent non-technical discussion of global warming, what needs to be done, and what is being done, see Walker and King (2008). For a grimmer picture, see Orr (2009). For a list of recent books on global warming see:
http://www.kings.cam.ac.uk/sites/default/files/library/global-warming-bibliography.doc.

nuclear. One imagines efforts in this direction are even now underway. It is thus a matter of some considerable importance that the threat of terrorism is combated intelligently, in such a way as to give some hope that the threat will decrease. Elsewhere,[29] I have argued that, in combating international terrorism, eight basic principles ought to be observed:-

1. International law must be complied with.
2. Terrorism must be combated as a police operation, not a war.
3. Civil liberties must not be undermined.
4. Nations suspected of harbouring or supporting terrorists must be engaged with both by means of diplomacy, and in such a way that intelligence is sought by stealth.
5. If terrorists' acts are motivated by long-standing conflict – as in the Palestine/Israeli conflict – every effort should be made by the international community of nations to resolve the conflict that fuels the terrorism.
6. As far as possible, terrorism must not be combated in such a way as to recruit terrorists.
7. International treatises designed to curtail the spread of terrorist materials must be maintained and strengthened.
8. Democratic nations combating terrorism must exercise care that, in combating terrorism, they do not thereby act as terrorists.

Unfortunately, the "war against terrorism", fought by George Bush, Tony Blair and others after 9/11, has violated all of these eight principles. As a result, the very opposite of what was intended – or at least ostensibly intended – has been achieved. The threat of terrorism has increased. Failure to resolve the Israel/Palestine conflict, unjust imprisonment, torture, the rhetoric of war and conflict of civilizations, and above all, perhaps, the Iraq war, have provoked young Islamic men all over the world to become sympathetic to bin Laden's cause. The response to

---

[29] See my 'The Disastrous War against Terrorism: Violence versus Enlightenment', Ch. 3 of *Terrorism Issues: Threat Assessment, Consequences and Prevention*, ed. A. W. Merkidze, Nova Science Publishers, New York, 2007, pp. 111-133 (available on my website www.nick-maxwell.demon.co.uk).

terrorism may indeed be judged to be more dreadful than the terrorism itself – especially when one takes into account that, at the time of writing, something like 1.2 million people have died as a result of the Iraq war, according to one estimate.

One has to remember that both Bush and Blair were re-elected after it had become apparent what a disaster the Iraq war had been. Bush and Blair would not have been able to conduct their disastrous war against terrorism without considerable popular support. A prerequisite, for dealing with terrorism intelligently and effectively, in other words, is a public with enlightened ideas about what kind of strategies are most likely to meet with success. Solving the problem of terrorism – in so far as it can be solved – requires public education.

Wisdom-inquiry is designed to provide just such an education, whereas knowledge-inquiry is not. What is needed is enlightened public debate about what needs to be done to combat terrorism successfully. It is just this which wisdom-inquiry would seek to provoke. Academics working in accordance with the edicts of knowledge-inquiry may well set about acquiring expert knowledge about terrorist groups, but it is not a part of their professional brief to provoke public debate about anti-terrorist policy.[30]

Global warming and terrorism illustrate the general point. As I have already remarked, our only hope of tackling our global problems successfully is to tackle them *democratically*. This in turn requires that electorates understand what these problems are, and what needs to be done in response to them. And this in turn requires that we possess institutions of learning rationally designed, *well* designed, to help the public discover what our

---

[30] For an excellent discussion of how to combat terrorism see: L. Richardson, 2006, *What Terrorists Want*, John Murray, London. This seems to me exactly the kind of work that academics today should be writing: intelligent, informative, wise, highly readable and well-written, it provides genuine insight into the motives and character of terrorism, and comes up with sensible proposals as to how the problem should be tackled. It is clearly intended to contribute to public education. It is an exemplary contribution to wisdom-inquiry. Some wisdom-inquiry work does go on in universities today despite the dominance of knowledge-inquiry intellectual standards.

problems are, and what needs to be done in response to them. Wisdom-inquiry is designed to provide this vital kind of education about our problems of living, while knowledge-inquiry is not. Wisdom-inquiry is designed to help humanity come to grips with the fundamental problem of enabling the God-of-Cosmic-Value to flourish within the God-of-Cosmic-Power in all its detailed, multitudinous aspects. Knowledge-inquiry violates three of the four most elementary rules of reason when judged from this standpoint.

In subsequent chapters, I will strengthen the argument that has, so far, only been lightly sketched; and I will indicate just how widespread are the damaging repercussions of failing to develop institutions of learning well designed from the standpoint of helping the God-of-Value to flourish within the God-of-Power.

A New Approach to Philosophy?
The subtitle of this book requires, perhaps, a few words of explanation. For the last fifty years, academic philosophy has been divided into two schools: analytic philosophy, based primarily in the USA and UK, and continental philosophy, based in Europe but to be found in the USA and UK as well. Analytic philosophy is based on the idea that the proper task of philosophy is to analyse concepts and solve puzzles. It seeks to clarify such key notions as knowledge, mind, justice, the good, reason, truth. Analytic philosophers strive for clarity, respect science and logic, and seek to support philosophical views with valid argument. Continental philosophy is an admixture of such schools as phenomenalism, existentialism, structuralism and Marxist critical theory. Science and valid argument are not high on the agenda. It tends to be bombastic and obscure.

Both, in my view, miss the point. Both fail to do philosophy in the way in which it most needs to be done. Philosophy ought to be the enterprise of tackling our most general, fundamental, urgent problems. The basic task of philosophy is to articulate, and try to improve the articulation of, our most fundamental problems and, if possible, propose and critically assess possible solutions. This should be done in as simple and non-technical a way as possible. A major task for philosophy is to keep alive an awareness of what our fundamental problems are, the relative inadequacy of our attempted solutions, and the impact that the answers we give or assume can have on diverse aspects of life – politics, science, art, education, religion, international affairs, the law, personal life, even survival.

It is philosophy in this sense that I attempt in this book.[31]

---

[31] It could be said that this is not a *new* approach to philosophy, but rather a return to a very old approach – philosophy as pursued by those who invented it, the ancient Greeks. Thales, Anaximander, Parmenides, Heraclitus, Xenophanes, Democritus, Socrates, Plato and Aristotle all did philosophy in the sense I have indicated.

CHAPTER THREE

HOW IS IT POSSIBLE FOR THE GOD-OF-VALUE TO EXIST INSIDE THE GOD-OF-POWER?

How is it possible for the God-of-Cosmic-Value to be inside the God-of-Cosmic-Power? How can our human world as we experience it, full of sound, colour, feel, taste and smell, imbued with sentience, consciousness, free will, meaning and value, containing everything we love and care for, this whole experienced world in which we live, somehow be embedded in, or be an integral part of, the physical universe?

I sit in my garden with a friend. Behind me, honeysuckle tumbles over the garden wall, and fills the air with its sweet scent. Bumble bees buzz and blunder among the honeysuckle flowers. A gentle breeze sifts through the tree above, and sunlight filters through the leaves. It is summer. The sky is dark blue. I stretch and say "This is heaven," and my friend replies "How right it is to take the garden as an image of Paradise".

Put all this into the physical universe and what do we have? Both I and my friend seem to disappear altogether. I am made up entirely of billions of cells, which are in turn made up of billions of highly complex molecules, in turn made up of atoms, in turn made up of electrons, protons and neutrons, the protons and neutrons in turn made up of quarks. Everything I am, everything I do, think, experience, see, feel, imagine, decide, understand is just billions of electrons and quarks interacting with each other in accordance with the laws of physics. And likewise for my friend. I see the blue sky, the green leaves, flowers and ferns; I smell the honeysuckle, and hear bees buzzing, and say "This is heaven." But what has really happened? Light of various wavelengths, reflected from various surfaces, enters my eyes where it causes molecular processes to occur in my optic nerves; this in turn causes more such molecular processes to occur in the back of my brain, which lead to more such processes which, in turn, lead to muscles being contracted, air being expelled, vocal chords vibrating, vibrations of molecules in the air, which cause my friend's eardrums to vibrate, in turn causing tiny bones in her middle ear to vibrate, leading to complex molecular processes in her brain. Ultimately, all that has occurred is that billions upon billions of electrons and quarks interacting with one another have produced light of such and such frequency which, after travelling short distances, have affected the way further billions of electrons and quarks interacted. Colours disappear; sounds and smells disappear; perceptions, experiences, sensations, feelings, consciousness, intentions, decisions and actions disappear, and there remains merely electrons and quarks interacting, these interactions being mediated by forces such as electromagnetism, the weak and strong force, and gravitation, vibrations in the electromagnetic force travelling from one vast conglomeration of electrons and quarks to another. All meaning and value, everything necessary to have anything meaningful or of value, have vanished, leaving only cold physics behind. The God-of-Cosmic-Value vanishes in the cold embrace of the God-of-Cosmic-Power.

How is our precious human world to be rescued from this insidious and terrifying assault from physics? The essential step is to recognize that physics covers everything, but only a highly selected *aspect* of everything. Physics, and that part of science in principle reducible to physics, is concerned only with what may be called the "causally efficacious" aspect of things, that aspect which, ultimately, everything has in common with everything else, and which determines (perhaps probabilistically) the way events unfold.[1] In its almost dementedly single-minded determination to specify precisely the causally efficacious aspect of things, physics ignores entirely all other aspects - the look of things, the feel, the smell, the sound, the sense, and what it is to *be* such and such a complex mass of cells, of interacting molecules. Physics fastens

---

[1] The thesis that physics seeks to specify the *causally efficacious* aspect of things is expounded and defended in my (1968a), reprinted in Swinburne (1974), pp. 149-74. See also my (1998), especially pp. 141-55.

onto the wavelength of light and ignores its colour; it specifies vibrations in the air and ignores the sound of the human voice, and ignores, too, what the person says. Physics might cover all the incredibly complex *physical* processes going on inside my head,[2] but it says nothing about what it is to *be* me, what it is I experience, feel, think, see, hear, imagine, understand, desire, fear, intend, decide. Given my utterance "This is heaven", physics covers comprehensively the causally efficacious aspect of what goes on – neurons firing in my brain, muscles contracting, vocal chords vibrating, vibrating molecules of the air, vibrating bones in the middle ear, more neurons firing in my friend's brain. In its almost monomaniacal concentration on the causally efficacious, however, physics leaves out entirely what the utterance sounds

---

[2] It is important to appreciate that physics can, in practice, predict the evolution of only the very simplest of systems. This is partly because the instantaneous initial state of the system – the so-called "initial conditions" – can only be specified for the very simplest of systems (and inevitably with some imprecision). It is also because it is in practice only possible to solve the equations of the physical theory – quantum theory, for example – for the very simplest of systems. Quantum theory can be applied with some precision to an isolated hydrogen atom, the very simplest atom in existence consisting of one proton and one electron. But when it comes to giving a precise quantum mechanical description of the next simplest atom, a helium atom, consisting of a central nucleus of two protons and two neutrons encircled by two electrons, it turns out to be impossible to give precise solutions to the equations of the theory. The best that can be done is give approximate solutions, and thus an approximate quantum mechanical description of the physical state of the atom. More complex isolated molecules would pose even more severe, if not insuperable problems. Giving a quantum mechanical description of a single neuron will, for ever, be beyond the scope of physics. A precise quantum mechanical description of a living, conscious brain is absolutely out the question. But these are practical obstacles. The physical state of a conscious brain and its environment at an instant, plus the yet-to-be-discovered true physical theory of everything (assuming it exists), does determine (perhaps probabilistically) the next physical state of affairs, even if human physicists are unable to carry out the calculations and derivations.

like, what it means, what I intend it to mean, what my friend hears and understands.

The Two Aspect View

This solution to the problem of how the God-of-Value can fit inside the God-of-Cosmic-Power is a new version of an old idea, which goes back at least to Spinoza, and is sometimes called the *two-aspect theory*. According to this view, there are two aspects to what exists, the physical on the one hand, and the mental, experiential, evaluative or human on the other. Everything that exists has a physical aspect. Some things also have experiential or human aspects. Thus flowers have colours and smells. People have thoughts, feelings, inner experiences; they have personalities imbued with features of value, such as courage, meanness or kindness, and live lives more or less meaningful and of value. Books contain sentences that make sense, and tell stories or expound ideas; and works of art are imbued with aesthetic features. None of these experiential or human features is reducible to the physical.

A simple argument, usually attributed to Thomas Nagel[3] and Frank Jackson[4] (but actually first spelled out by me several years earlier[5]) establishes that an elementary experiential property, such as *redness*, cannot be reduced to the physical. Being blind from birth does not debar one from understanding the whole of physics. A person blind from birth may well be able to understand the physical theory of light as well as the next person. Physical concepts such as wavelength, mass, charge, force, momentum, are such that no special kind of experience is necessary for their meaning to be understood. But when it comes to sensory qualities, such as colours, sounds and smells, the situation is quite different.

---

[3] Nagel (1974).
[4] See Jackson (1982) and (1986).
[5] See my (1966), especially pp. 303-308; and my (1968b), especially p. 127, pp. 134-137 and 140-141. When I drew Thomas Nagel's attention to these publications, he remarked in a letter, with great generosity: "There is no justice. No, I was unaware of your papers, which made the central point before anyone else". Frank Jackson acknowledged, however, that he had read my 1968b paper.

Here, it is necessary to have experienced these sensory qualities oneself, at some time in one's life, to know what they are. Being blind from birth does debar one from knowing what redness is, the colour we see and experience. In short, being congenitally blind does not debar you from knowing and understanding everything that can be predicted by physics, but it does debar you from knowing or understanding what it is for a poppy to be red; hence the redness of the poppy cannot be predicted by physics. Colours, sounds, smells, tactile qualities, sensations, feelings, thoughts, and a multitude of other experiential, human features of things and people, lie irredeemably beyond the scope of physics.

But might not some future development of current physics successfully predict and explain colours, sounds and smells as we experience them? The answer is No. Suppose physics one day completes its task of discovering the true "theory of everything" which, in principle, predicts and explains all physical phenomena. Given any isolated system, this theory, together with a precise specification of the instantaneous physical state of the system, would (in principle) predict all future states of the system, as long as the system remains isolated.[6] Such a theory would clearly be complete and comprehensive in a dramatic and extraordinary way.

But, despite this, the theory might well not predict all facts about a system. It predicts only those facts that need to be specified (at any given instant) for further predictions to be made. All facts and properties which do not need to be referred to for the above kind of predictive task to go ahead, are ruthlessly excluded from physical descriptions.

Thus, suppose the isolated system is a space capsule with you inside. The physical state of the capsule, and the physical state of your brain and body, are included in any complete specification of the physical state of the system, used to predict future states, described in similar terms. But colours and sounds that you experience, your inner sensations and thoughts, the meaning of

---

[6] There are, however, severe restrictions on what physics can predict *in practice* as I pointed out in note 2. In the text, for simplicity, I assume determinism. The true physical "theory of everything" may, however, be probabilistic, in which case probabilistic predictions only would be forthcoming, even in principle.

what you say or write down in your diary, are all excluded from the physical description because these are not required for the predictive task of physics to go ahead. Physical aspects of these experiential, human features are, of course, specified: light of such and such a range of wavelengths, sound waves in the air, neurological processes in the your brain, ink marks in your diary – all are specified in terms of the instantaneous states of fundamental physical entities. But the experiential, human aspects of these physical processes receive no mention, because they are not required for the predictive task of physics to succeed.

But could one not develop an even more comprehensive theory than the physical "theory of everything", by adding on additional postulates which correlate physical states of affairs with experiential features – redness, the sensation of redness, and so on? This new theory would be really complete and comprehensive: it would predict everything, the physical *and* the experiential.

But a terrible price would be paid. The new theory would not be explanatory. In turning the *physical* "theory of everything" into a *real* theory of everything, one would have to add on endlessly many postulates linking the physical and the experiential, each one of which would be incredibly complex. Even the postulate linking physical states of affairs to a particular hue of red would be extraordinarily complex. The number of such incredibly complex additional postulates is endless, as becomes apparent when one considers the diversity and richness of our human experiential world, and adds on to that the experiences of other sentient creatures, actual and possible. The *physical* theory of everything will be explanatory because, like existing physical theories, it will have an extraordinarily simple, unified basic structure. The *real* theory of everything will, by contrast, have billions, possibly even infinitely many, distinct postulates, each one of horrendous complexity. Such a theory might predict, but it would be hopelessly non-explanatory.

Two crucial points emerge from these considerations.

First, physics does, in a very important sense, seek to be comprehensive and complete. It seeks to develop that unique true, unified "theory of everything" which, in principle, applies to all possible phenomena and which, in principle, predicts future states of any isolated system *when specified in terms of the vocabulary of the theory, in those terms required in order to predict further states of the system.* Physics is, in other words, as I have said, exclusively concerned with the *causally efficacious* aspect of things, that aspect which, ultimately, everything has in common with everything else, and which must be specified if physics is to succeed with its predictive task. All those aspects of things which do not need to be specified for the predictive task of physics to succeed are not specified. They are ruthlessly excluded from physics. The look of things, the feel, sound and smell of things, what it is to *be* such and such brain and body, the meaning and value of things: all these experiential and human features of things are excluded from physics because no mention of them need be made in order to fulfil the predictive task of physics.

Second, there is an *explanation* as to why physics must be silent about the experiential. Leaving out the experiential from physics is the price that must be paid if the beautifully explanatory theories of physics are to be developed. Physics must be silent about the experiential, not because it does not exist, but because bringing in the experiential destroys utterly the explanatory character of physics.

In short, physics omits all references to colours, sounds and smells as we experience them, not because they do not exist, but because (1) physics *can* omit all reference to them without this sabotaging its basic predictive task, and (2) physics *must* omit all reference to them if physical theory is to be explanatory.

A key point in all this is that the silence of physics about sensory qualities, inner experiences, meaning and value, our whole experienced human world (the God-of-Cosmic-Value) provides no grounds whatsoever for holding that these features of things do not exist. A comprehensive physical specification of the God-of-Cosmic-Power says nothing about the God-of-Cosmic-Value, not because the latter does not exist, but because prediction does not require it, and explanation demands that it be omitted.

The Orthodox Scientific View
Almost all scientists, however, have reached the opposite conclusion. They take it for granted that the silence of physics

about the experiential means that sensory qualities do not exist, objectively, out there in the world and, at best, exist only in us, as sensations. According to this view, the whole world as we experience it is nothing more than a persistent illusion. The blue of the sky, the green of leaves and grass, the yellow of honeysuckle flowers, the buzzing of bumble bees, the sounds of conversation, all this is illusory. Out there, there is just physics: invisible fundamental physical entities, electrons, quarks, photons and so on, interacting with incredible rapidity in accordance with the laws of quantum theory (or, more precisely, the laws of the true "theory of everything"). We have no experience of this physical world whatsoever. Everything we do experience – everything we see, hear, touch, smell – is in us, not out there in the world. The world outside us *causes* us to have the experiences we do have, but we never directly experience these external causes, and what we do experience provides us with a vivid but almost wholly misleading impression of what these external causes are.

This view goes all the way back to Democritus, over two thousand years ago, one of the first to conceive of the world in purely physical terms. Democritus held that the universe is made up exclusively of indestructible atoms which move through the void. And he declared:

> Colour exists by convention; sweet and sour exist by convention: atoms and the void alone exist in reality.[7]

Two thousand years later, in 1632, Galileo expresses the same view:

> these tastes, odours, colours, etc., so far as their objective existence is concerned, are nothing but mere names for something which resides exclusively in our sensitive body, so that if the perceiving creatures were removed, all of these qualities would be annihilated and abolished from existence.[8]

---

[7] A slightly different translation is quoted in Guthrie (1978), p. 440.

[8] Galileo, *The Assayer*, quoted in Matthews (1989), pp. 56-7.

Galileo goes on to point out that if a feather tickles us we hold that the tickling is in us, not in the feather. In a similar way, colours, sounds and smells are a kind of tickling in us, and are not objective features of things external to us.

And Newton agrees. He writes:

> if at any time I speak of light and rays as coloured or endued with colours, I would be understood to speak not philosophically and properly, but grossly, and accordingly to such conceptions as vulgar people in seeing all these experiments would be apt to frame. For the rays to speak properly are not coloured. In them there is nothing else than a certain power and disposition to stir up a sensation of this or that colour. For as sound in a bell or musical string, or other sounding body, is nothing but a trembling motion, and in the air nothing but that motion propagated from the object, and in the sensorium 'tis a sense of that motion under the form of sound; so colours in the object are nothing but a disposition to reflect this or that sort of rays more copiously than the rest; in the rays they are nothing but their dispositions to propagate this or that motion into the sensorium, and in the sensorium they are sensations of those motions under the forms of colours.[9]

Almost all scientists today would agree. Thus Semir Zeki, a present day neuroscientist who has done much to unravel the neurology of colour perception, writes "Ever since the time of Newton, physicists have emphasized that light itself, consisting of electromagnetic radiation, has no colour"; and Zeki goes on to quote a part of the above passage from Newton with approval.[10]

If the only reason for holding that sensory qualities, as we perceive them, do not exist out there in the world is that physics is silent about them, one might dismiss the views of these scientists as being based on nothing more than the unconscious assumption that nothing can in principle elude the grasp of physics. They

---

[9] Newton (1932), pp. 124-5.
[10] Zeki (1993), p. 238.

assume that the silence of physics about colour as we experience it must mean that colour does not objectively exist, ignoring the possibility that physics is silent about colour because it is not needed for prediction, and it cannot be included if physics is to be explanatory.

But there is another reason for believing sensory qualities do not exist out there in the world – a reason that may well seem decisive.

It might be thought somewhat paradoxical that *scientists*, of all people, should deny the objective existence of colour, sound, smell, tactile qualities, as perceived by us. For consider the theory that the material world is made up of physical entities that are devoid of colour, sound and smell. Is not this theory refuted by experience the moment we open our eyes, our ears, sniff and put out a hand and touch? The world around us, as most of us experience it, is full of colour and sound, smells and tactile qualities. Scientists above all claim to base their science on experience, on evidence, on observation and experiment. They above all, surely, should acknowledge that the theory of the sensory-depleted world is refuted by every trivial observation that we make.

But there is an immediate and apparently devastating reply to this objection, this claimed refutation. Take the sensory-depleted physical conception of the universe, as a working hypothesis, and apply it to what goes on when we see, hear, smell, feel. I see before me the glistening bank of golden honeysuckle flowers in my garden. Sunlight containing a range of wavelengths strikes the molecules of which the honeysuckle flowers are composed: some wavelengths are absorbed, others are reflected. Some of this reflected light enters my eyes. It is focused onto the retinas of my eyes, where it causes complex chemical processes to occur in the receptor cells of my eyes, which in turn cause ripples of neurological activity to travel along the optic nerves to the back of my brain. This neurological activity takes the form of a rapid exchange of sodium and potassium ions across the semi-permeable membrane of the neurons that make up my optic nerve, this exchange travelling as a wave of activity from the retina to my brain. Further such neurological activity goes on at the back of my brain – and then the miracle occurs: I have the experience of seeing

62

the yellow honeysuckles. What I know about, here, is the last event in this long chain of events: the experience of seeing apparently golden honeysuckle flowers. But this experience is inside me, and is utterly different from its external cause. Indeed, science tells us that my inner experience of seeing is the outcome of a series of dramatic transformations: reflected light becomes chemical processes in the receptor cells of my retinas, which in turn become neuronal impulses travelling along my optic nerve, in turn transformed to the transmission of particles at synaptic junctions, all this brain activity being transformed into my private experience of seeing. This final occurrence, which is all I really know about in seeing, is wholly different from what has gone before – brain activity, chemical activity in my retinas, electromagnetic radiation absorbed and reflected by molecules that make up the physical honeysuckle flowers. Thus perception provides me with no reason whatsoever to suppose that things external to me really do have the perceptual properties that they appear to have. On the contrary, the above sketchy scientific account of what goes on during perception provides us with every reason to suppose that things external to us are utterly different from the way they appear to be to us when we see them.

Similar considerations arise in connection with hearing, smelling, and touching. And the conclusion is that the scientific, sensory-depleted theory of the universe is not refuted at all by our ordinary perception of colour, sound, smell and tactile qualities. Quite to the contrary, science provides us with every reason to hold that things out there, the world around us, is utterly different from the way it seems to be when we see, listen, sniff, touch. Galileo was right. These sensory qualities are in us, not in things around us.

Cartesian Dualism

We have arrived at an immensely influential attempted solution to our problem, formulated very clearly by Rene Descartes[11] around 1656, and known ever since as "Cartesian dualism". According to Cartesian dualism, there are two universes – the

---

[11] Descartes (1949).

physical universe on the one hand, and the universe of conscious minds on the other. Each living human brain has associated with it a conscious mind, and everything we see, hear, feel, experience is confined to our conscious mind. The God-of-Cosmic-Power is the physical universe; the God-of-Cosmic-Value is to be associated with distinct conscious minds, linked to but distinct from our physical brains. The strongest argument in support of Cartesian dualism is the one just given above, from physics and the complex processes associated with perception

Persuasive as this argument for Cartesian dualism may seem, it can nevertheless be challenged. There are grounds for doubting that it really is valid.

First, one should note that the conclusion of the argument has some awkward consequences. If what I really *see* when I perceive the honeysuckle – if what I really know about – is my inner experience of seeing, and not the honeysuckle at all, then this must mean that there is this inner mental representation of the honeysuckle, just as it appears to me, associated somehow with neurological processes going on in my brain. In so far as I know anything about anything, I know that this inner mental representation of honeysuckle exists, and I know – I *see* – what its properties are: the honeysuckle shape, colour and arrangement, but somehow in my private mental space, and not in ordinary physical space. But this inner mental honeysuckle representation is a very peculiar sort of entity indeed. Only I am aware of it; no one else can detect the faintest hint of its existence, however exhaustively my conscious brain might be examined. It is an entity with properties that differ radically from anything known to be going on in my brain. It is utterly obscure how neurological processes occurring in my brain manage to create this weird mental entity – manage to cause it to occur. And furthermore, if all I ever know about, when I see, hear, smell and touch, are my inner mental representations, how can I ever know anything else? How can I know anything about the world around me? All these awkward consequences flow from the conclusion that what I really see, what I really know about, when I look at honeysuckle, is my inner visual experience, and not the external honeysuckle. We may well hold that Cartesian dualism creates more problems than it solves.

But how valid is the above argument that, given the long chain of events stretching from external object to internal representation, what we really *see* and know about is the inner representation and not the external object? What prevents us from taking exactly the opposite view: what we really see and know about is the external object, and not our inner representation at all?

Internalism versus Externalism

We have here two radically different theories about what it is that we really see and know about in perception. The "internal" theory – as we may call it – associated with Cartesian dualism, says that what we really see and know about is our inner experiences, our inner representations of external objects. Whatever we know about external objects is inferred from our primary knowledge of our inner experiences, our inner representations. The "external" theory says exactly the opposite. What we really see and know about is what we ordinarily take ourselves to see and know about: things external to us, flowers, people, houses, trees. We never *see* our inner experiences at all. All our knowledge of our inner experiential representations is inferred from our knowledge of things external to us. The "internal" theory says we know everything about our inner perceptual experiences and hardly anything at all – at least in ordinary perception – about things external to us. The "external" theory says the exact opposite: in so far as we know anything when we perceive, what we know about is things external to us; we hardly know anything at all about the real nature of our inner perceptual experiences, and what we do know is derived from our knowledge of things external to us.

The "external" theory[12] acknowledges the existence of the

---

[12] The distinction being drawn here, between the "internal" and "external" views, should not be identified with distinctions that have been drawn by other philosophers between "internalist" and "externalist" views. The Stanford Encyclopedia of Philosophy has 73 items under the heading "Externalism", involving many different distinctions between "internal" and "external" views, some of which are associated with inner experiences: see http://plato.stanford.edu/search/searcher.py?query= externalism. The distinction I draw here should be understood in the

causal chain of events from perceived external object to inner perceptual experience, and declares that the existence of this chain of events provides no grounds whatsoever for holding that what we *really* see and know about is the last event in this chain – our inner perceptual experience. And when put baldly like that, the declaration seems valid. Why should the existence of the chain of events imply that we only really see the last event in the chain?

This only becomes plausible if *additional* assumptions are made.

We may assume, for example, that we only really see and know about what we are in direct contact with. Naively, before we learn some science, we may think that we really see objects external to us because our eyes send out our gaze to touch and be intimately in contact with these external objects. Then we learn that there is no such thing as a "gaze" emitted from our eyes. On the contrary, it is all the other way round: light, reflected from the external object, enters our eyes and is focused onto our retinas. What we really see, then, is the image on our retina (which, rather confusingly, is upside down!). But then it occurs to us that we are not in direct contact with our retinas either: chemical processes going on there cause neurological processes to occur in our optic nerve and brain, which in turn lead to the occurrence of our inner visual experience, and it is only *this* that we are in direct contact with, this alone being what we really see.

But what needs to be challenged about this argument – this remorseless process of driving the seen object into the eye and through the eye socket into the brain and into the mind – is the driving assumption that we only *really* see and know about what we are in direct contact with. This is just false. It is based on a very primitive and false theory of perception – the "gaze" theory – which is in turn based on the analogy with touching. According to this primitive view, we only really get to know things external to us by reaching out and getting hold of things with our hands (perhaps even putting them into our mouths, as babies do). Seeing – gazing on things – is a kind of invisible touching which, from

---

way I indicate, and should not be interpreted in terms of the views of other philosophers. In so far as it owes anything to another philosopher, it comes from J. J. C. Smart, as I indicate in the next note.

pressure from science, is driven to accept that we only touch, and hence only really see, not external objects, not images on our retinas, but our inner perceptual experiences. But all this is nonsense. Seeing is not any kind of touching. In order to see something, we don't have to be in direct contact with that thing. (And this "touching" theory of seeing doesn't even work for touching itself. For, of course, when we touch and feel something with our finger tips, sensory cells cause neurons to fire, which in turn cause neuronal activity to occur in our brain which in turn leads to our experience of feeling something. We are never in direct contact with what we touch.)

Reject the false assumption that we only really see and know about what we are in direct contact with, and the argument from the causal chain associated with perception collapses. The argument provides no grounds whatsoever for holding that we really see, not external objects, but internal representations of them.

But there is another assumption that we may make which, if true, would render the causal chain argument convincing. We may assume that what we really see is something about which we cannot be mistaken. If we really see something, then we know we are seeing it, and we know what sort of thing it is. It follows that I cannot really see the honeysuckle in my garden. There is always the possibility that I am hallucinating. It is for me just as if I am in my garden: I hear the buzzing bumble bees, I see the honeysuckle and smell their sweet scent. But actually I am in a laboratory: fiendish neurologists have stuck electric probes into my brain and are stimulating it to produce just the neurological processes that would occur in my brain were I to be in my garden. The result is that I seem to see the honeysuckle and hear the bees; but actually I see and hear nothing of the kind. From the standpoint of what I experience there is no difference whatsoever between being in the garden, and being in the fiendish neurologists' laboratory with the probes stuck into my brain. What I really see must be the same in both cases. Hence what I *really see* is my inner representation of the garden, the honeysuckle and the bees, and *not* these things themselves. Even when I am in my garden, I really see, not the garden, but my inner representation of it.

But the key assumption, here – that what we really see is something about which we cannot be mistaken – is *false*, and deserves to be rejected. All perception has a conjectural element to it. There is always the possibility – however remote – that we are mistaken about what we are seeing. There is always the possibility that we seem to be seeing something, and it does not exist at all.

What do I see, then, if I hallucinate that I am seeing honeysuckle, and nothing out there corresponds to what I seem to be seeing? In these circumstances I see nothing. If I am to see honeysuckle, (1) the honeysuckle must exist before me, (2) light reflected from the honeysuckle must enter my eyes and lead to me (3) having the experience of seeing the honeysuckle. If (1) and (2) are absent and only (3) remains, I see nothing.

But if this is the case, (3) alone holds, and I experience a hallucination, and know it is a hallucination, do I not know *something* as a result of my experience? The answer is that I *do* know something, but what I know is very little. What I know can be put like this: "Something is going on inside me which is just the sort of thing that would go on if I was really seeing honeysuckle".[13] In other words, I ordinarily know very little about the perceptual experiences I have; what I do know is derived from my knowledge of things external to me – what these perceptual experiences are experiences *of*. Far from my knowledge of things external to me being inferred from my knowledge of my inner experiences, it is all the other way round: all that I know of my inner perceptual experiences is inferred from my knowledge of things external to me. Even though I am in intimate contact with my inner perceptual experiences, my knowledge of their nature is sparse indeed.

This feature of the "external" theory means that this theory is free of the serious problems that confront the "internal" theory, indicated above. According to the "internal" theory, when I am in my garden I directly *see* and know about, not the flowers, the trees, the sunshine, the sky, but my inner perceptual experiences of these things. What I know about these inner experiences tells me that

---

[13] This way of putting it is due to J. J. C. Smart: see his (1963), p.94.

they are utterly unlike anything going on in my brain. Nothing like the fluttering green leaves, the sunshine, the blue sky, as seen by me to be features of my private inner perceptual experience, exists in my brain. At once the problem of what can be the relationship between my inner experience and my brain takes on a particularly severe form, as we saw above. Adopt the "external" theory, however, which declares that it is the flowers, trees, sunshine and sky that I really see, my knowledge of my inner perceptual experiences being very sparse – being limited to "what is going inside me is just the sort of thing that would go on given I am in my garden seeing trees, sunshine, sky" – and the horrendous problem of the relationship between inner experience and brain all but disappears. All that I do know about my inner experiences – which is very little, and couched in terms of what I know of things external to me – is entirely compatible with the conjecture that my inner experiences are brain processes. Everything I know about my inner experiences, such as the circumstances when they occur, what they are of, fits in with this conjecture, and nothing that I know about them conflicts with this conjecture. Of course, in declaring inner experiences to be brain processes I am not saying that inner experiences are nothing but physical processes going on in my physical brain – any more than in declaring the tree in my garden to be a physical object (or persisting physical process) I am saying it is nothing more than a physical object (or process). Just as the tree has extra-physical sensory qualities associated with it – the greenness of the leaves, the sounds it makes when the leaves swish against each other in the breeze, even a faint leafy scent associated with the tree – so too brain processes have experiential features associated with them – what I experience and am aware of when these processes occur in my brain. (There will be more about this in chapter seven.)

There is a third consideration which may make the existence of the causal chain of events associated with perception seem powerful grounds for holding that what we *really see* are our inner perceptual experiences and not things external to us. Invoke the causal chain of events involved in perception – from light absorbed and reflected at the surface of an external object, via chemical processes occurring in the retina, to neurological processes going on in the brain – and one thereby automatically invokes physics, since physics is what is required to describe and explain this causal chain in full generality. As an immediate consequence, all sensory qualities, such as colours, disappear from the external world. We have, however, only limited scientific knowledge and understanding of our incredibly complex brains; inevitably, there is the temptation to suppose that in this region of our ignorance something as yet not understood and mysterious happens, and brain processes cause our inner perceptual experiences to occur, we experience colours and sounds which we, mistakenly, project outwards onto things in the world around us.

But there is a double fallacy here. First, as we have seen, the silence of physics about colours and sounds in the world around us provides no reasons whatsoever for thinking that these sensory qualities don't exist. If they do exist, physics would not mention them (since they are not required for physical prediction, and cannot be included if we are to have physical explanation). Second, the reasons for physics to be silent about experiential qualities in the world around us, are precisely also reasons for physics to be silent about the experiential aspects of our inner experiences. Once we enter into the world of physical descriptions and explanations, we will never encounter either the greenness of leaves or the inner perceptual experience of seeing green leaves: instead we encounter electromagnetic waves of various wavelengths being absorbed and reflected by the molecules that go to make up the leaves' surfaces, and neurological processes going on in our brains described as physical processes. If we hold that colours and sounds don't really exist in the world around us, then we should also hold that the extra-physical, experiential aspects of our brain processes, our states of awareness, don't really exist either. The arguments for the non-existence of the one are as good – or as bad – as the non-existence of the other. The complexity of the brain does not provide an honourable shelter to resist the import of this argument.

Do Sensory Qualities Exist Objectively?

But still – it may be protested – surely colours, sounds and smells as (most of us) experience them do not exist *objectively* out

70

there in the world around us. These sensory qualities are surely *subjective*, dependent upon us: and to this extent Galileo was right when he said that they exist in us, so that if all sentient creatures are removed, all sensory qualities would vanish from the world.

It all depends on what you mean by "objective" and "subjective".

You may mean that if something exists it is "objective" whereas if it only appears to exist but actually does not exist then it is subjective. My claim is that colours, sounds and smells really do exist out there in the world: in terms of this first distinction, then, colours, sounds and smells are *objective* (except when, as in illusions, after-images and hallucinations, they are *subjective*).

On the other hand you may mean that a property is independent of and unrelated to us, then it is *objective*, whereas if it is dependent on or specially related to us, then it is *subjective*. In so far as physical terms such as mass, electric charge, momentum, refer to real physical properties whose existence has nothing to do with human beings at all, then these properties are, in this second sense, *objective*. But a property like *poisonous*, which does relate quite specifically to us, to our physiology, is *subjective*. Colours, sounds and smells as we experience them are also *subjective*, since these properties relate quite specifically to our idiosyncratic physiology, our particular sense organs and brains. I assume here that in order to experience colours, sounds and smells as we experience them a conscious being, an alien perhaps, has to have sense organs and brains sufficiently similar to ours; there is simply no other way in which these qualities can be known about.

In one sense, then, colours are *objective*. In another sense, they are *subjective*. The crucial point, however, is that in declaring colours to be subjective in this second sense, one is not declaring that they don't really exist out there in the world, any more than that, in declaring the poisonousness of arsenic to be *subjective* in this second sense, one is declaring that this poisonousness does not really exist out there in the world. Being subjective, in the second sense, means that the property in question relates specifically to us human beings: aliens may well not find leaves green, or arsenic poisonous. But that does not mean that these *subjective* properties, in the second sense, do not really *objectively* exist, out there in the world around us – in the first sense of *objective*. Colours, sounds and smells (as we experience them), and poisons, are *objective* in the first sense (in that they exist out there in the world), but *subjective* in the second sense (in that they relate specifically to us, and may be indiscernible and unknowable to aliens with physiologies different from ours).

Failure to distinguish these two kinds of meaning that can be given to *objective* and *subjective* is perhaps responsible, more than any other factor, for the view that out there physics reigns supreme, and colours, sounds and smells, and the whole of the God-of-Cosmic-Value, have no kind of objective existence in the world. The sensible conclusion is reached that colours, sounds and smells are *subjective* in the second sense; failure to distinguish the two senses then forces one to conclude that colours, etc., must also be *subjective* in the first sense. One is thus forced to conclude that colours don't exist out there in the world at all.

One implication of this whole line of thought is that the world is likely to be a much richer and stranger place than we might at first imagine. It is reasonable to suppose that sentient creatures can exist with sense organs and brains different from ours. Indeed, such creatures exist here on earth. Bats see – or rather hear – the world in a way that is quite different from the way we see it. Sentient and conscious creatures on other planets, if they exist, are likely to see perceptual qualities as unimaginable to us as colours are to those blind from birth. Even if such aliens do not exist, it may nevertheless be *possible* for them to exist, which is all that is required for alien perceptual qualities to exist (which would be perceived if these aliens did exist). All around us, then, there are alien perceptual qualities of which we can have no inkling – perceptual qualities that are *objective* in the first sense, but *subjective* in the second sense.

A final argument for the non-existence of these perceptual qualities (*objective* in the first sense, *subjective* in the second sense) deserves to be mentioned. If these perceptual qualities really do exist in the world around us, then there is here an impenetrable mystery for science. For these properties lie forever beyond the scope of scientific explanation. Rather than admit such inherently scientifically inexplicable properties, surely we should grant that they are illusory and do not really exist.

There are two points to note about this argument. First, it applies just as much to extra-physical features of our inner experiences as it does to sensory qualities of things in the world around us. If the argument provides good reasons for not believing in the existence of colours, sounds and smells in the world around us, then it also provides good reasons for not believing in the existence of extra-physical features of our inner experiences. But second, the argument does not provide good reasons for not believing in the existence of colours, etc. in the world around us. Physics is concerned exclusively with the *causally efficacious* aspect of things, that aspect which, ultimately, everything has in common with everything else and which determines (perhaps probabilistically) how events unfold. Other aspects, such as what things look like, sound like, smell like, feel like, what it is like to *be* the physical processes in question, what uttered sounds or typed script *mean*, the *content* of a novel, picture or symphony: all these aspects, having no causal import, are ignored by physics, do not need to be specified by physics and, incidentally, cannot be specified if physical theory is to retain its explanatory power. As we saw above, there is an *explanation* as to why physics does not, and cannot, encompass the experiential aspects of things. It is not as if there is a mystery here, which physics ought to be able to comprehend, and cannot. The false impression that there is a mystery here (or would be if colours, etc., existed) arises from an exaggerated idea of what physics seeks to achieve. It arises from taking too literally the aim of physics to provide a *complete and comprehensive* account of the features of things in the world. Physics is concerned to cover *everything that exists*: but it is concerned to capture only the *causally efficacious* aspect of everything that exists. All other aspects are outside its sphere of interest. That physics says nothing about these other aspects is not a *failing* of physics. It does not mean that these other aspects are inherently inexplicable. It just means that they are not required for physical prediction, and need to be ignored for the purposes of physical explanation.

What Do We Really See?

It should be noted that throughout the above somewhat intricate arguments I have not argued positively that colours, etc., as we experience them, really do exist out there in the world around us. Rather, I have demolished a series of arguments designed to establish that colours do not exist out there. Superficially it looks as if physics establishes the non-existence of colours. But this is a mistake. Physics establishes nothing of the kind.

There are two rival theories about what we really see in perception before us: the "internal" and "external" theories. Physics seems to favour the "internal" theory, but actually it does not. How, then, are we to decide between these two theories? What in general determines what we see? We have considered the chain of events that stretches from honeysuckle to my brain when I see, or seem to see, the honeysuckle. But the chain of events goes further back than this. It goes back, in part, to the surface of the sun, and further back, for millions of years into the past, into the interior of the sun. What is it that picks out a specific link, a specific place, in this long chain of events, to be that which I really *see*?

The answer is that we see what we ordinarily *know most about* as a result of a perceptual experience. In looking at the honeysuckle I don't see the interior of the sun because I know nothing about the interior of the sun as a result of the perceptual experience.

This answer, unfortunately, does not help much in deciding between the "internal" and "external" theories, because it is just on this matter of what we do know most about when we perceive that the two theories differ. How then are we to decide? There are four relevant considerations: all four favour the "external" theory.

First, as we have seen, the "internal" theory, as a result of holding that it is our internal perceptual experiences that we really know about when we see, hear, etc., thereby creates an enormous problem when it comes to understanding how these "internal perceptual experiences" are to be related to brain processes. As a result of insisting that we do really see, and thus know about, these internal perceptual experiences, the "internal" theory is committed to the existence of mental entities or processes with known mental features utterly unlike anything going on in our brains. It is utterly obscure what the relationship can be between brain processes and these mental processes. It is utterly obscure how brain processes can cause these mental processes to occur. And if what we know in perception is restricted to knowing about these internal mental representations of unperceived external causes, it becomes wholly obscure as to how we can know anything at all about the world external to our bodies around us.

The "external" theory faces none of these problems. For, according to this theory, we ordinarily know hardly anything about our inner perceptual experiences, and what we do know is inferred from our knowledge of things around us. We do not know enough about our inner experiences for this knowledge to go against the hypothesis that these experiences are brain processes, and what little we do know thoroughly fits in with this hypothesis (as long as it is recognized that brain processes have non-physical features or aspects). This difference strongly favours the "external" theory.

Second, the "internal" theory faces the severe problem of how we can ever know anything about anything other than our inner experiences. If what we directly perceive and know about in perception is our inner experiences, and all our knowledge of everything else is inferred from our knowledge of our inner experiences, what possible basis can there be for any knowledge at all of things external to us? We can never compare our inner representation of honeysuckle, let us say, and the honeysuckle itself, because we can never experience the external honeysuckle; we only know about our inner representations of it. All we can ever do is compare our own inner experiences, one with another, never the inner experience and what it is supposed to represent. Granted the "internal" theory it seems, in fact, that we cannot know anything about the external world, and cannot even know that it exists. Just this conclusion was reached long ago in 1709 by Bishop Berkeley.[14] Earlier, in 1690, John Locke defended what is usually called "the representational theory of perception"[15] – what I have called here the "internal" theory. Berkeley, taking this as

---

[14] Berkeley (1957).
[15] See Locke (1961), Book II.

his starting point, arrived at the conclusion that we only have knowledge of our inner experiences and can have no knowledge whatsoever of things external to us.

The "external" theory entirely avoids the problem. For, according to this view, what we really know about in perception is the external perceived object; our knowledge of our inner experiences is inferred from our knowledge of things external to us (the exact reverse of the "internalist" viewpoint). Berkeley's arguments do not establish idealism. Rather they amount to a *reductio ad absurdum* of the "internal" theory – Locke's representational theory of perception.

Third, one would think that evolution would arrange for animals (and for us) to get to know about relevant aspects of the environment in perception, aspects relevant to survival and reproduction, rather than arrange for animals (and us) to get to know about their (our) internal states and processes. Animals that find out about inner perceptual experiences, and not about the world around them, are not likely to survive for very long. They are not likely to evade predators, find food and mates, and care for offspring. And this consideration is reinforced by common sense: if we know everything about our inner experiences and hardly anything at all about the world around us, how can it be that we manage to act and survive as well as we do in the world? It is surely the basic biological function of perception to inform us about relevant aspects of the world around us. All this favours the "external" theory.

Fourth, there is the following consideration. Ordinarily it never occurs to us to doubt that we directly perceive things around us, and these things possess the perceptual properties we see them as having. (Occasionally we may suffer from illusions or hallucinations, but let's ignore these infrequent occurrences.) Along comes physics and seems to tell us that all this is a gigantic, persistent illusion. If the arguments from physics are valid, we must clearly revise our ideas about what we ordinarily see. But we only need to do this if the arguments from physics really are valid. If these arguments are all *invalid*, and provide not an iota of reason for abandoning our customary, common sense views, then we should hold onto these views, at least until something serious crops up to challenge them. But we have seen that all the arguments from physics *are invalid*. The most that they do is to revise our ideas about the objectivity of perceptual qualities to the extent of establishing that they are objective only in the first sense, and subjective in the second sense. Hence we should hold onto our common sense views about what it is that we see, hear, smell, touch.

The real reason for believing in the existence of sensory qualities as a part of the world around us – once the arguments from physics have been seen to be invalid – comes from our direct experience. We *see* the colours of the rainbow, we *hear* birds sing and the roar of traffic, we *smell* the sweet scent of honeysuckle and *feel* the smooth, hard surface of marble. We *experience* the God-of-Cosmic-Value, and that is our reason for believing It exists. It may seem that the God-of-Cosmic-Value, our rich, experienced world, withers and dies when put into the all-encompassing embrace of the physical universe, the God-of-Cosmic-Power. But this is a mistake. The God-of-Cosmic-Value and the God-of-Cosmic-Power can coexist.

The Mystery of Consciousness Solved?

The above considerations go a long way towards solving a major philosophical problem concerning the relationship between the mind and the brain. Ordinarily there may not seem to be anything very mysterious about our inner experiences – our perceptions, thoughts, imaginings, feelings, states of awareness. Invoke science to improve our understanding of these inner experiences, and the very opposite of what we may hope for occurs. Far from becoming more comprehensible, they vanish! View what goes on inside our heads through the lens of science, and instead of inner experiences, states of awareness, we see only the squashy brain, neurological processes, impulses travelling down millions of neurons and being transmitted across millions of synaptic junctions.   Our thoughts, feelings, imaginings, states of consciousness, far from becoming more comprehensible, have disappeared altogether. And what seems so disturbing about this annihilation of our inner being is that it occurs when we invoke our very best means for explaining and understanding things – namely

77

science. Far from rendering consciousness comprehensible, science annihilates it. The net effect is to make consciousness seem something that is utterly mysterious – so mysterious, indeed, that invoking our very best mode of explanation causes consciousness to evaporate, as if a mere will-o-the-wisp. All this prompts one to ask: What is consciousness? What are these mysterious inner experiences of ours? Why are they so elusive? What makes them disappear when one seeks to explain and understand them by means of science? And what is their relationship to that which science does reveal to us, namely the neurological processes going on in our brains?

These philosophical problems about consciousness have been solved by the considerations developed above. Consciousness evades scientific explanation because everything experiential evades science – or, at least, evades that part of science that is, in principle, reducible to physics. This, as we have seen, is the price that must be paid for physical theories to be explanatory. That consciousness seems to vanish when scientific explanation is invoked does not mean that there is something inherently mysterious or inexplicable about consciousness – any more than that colours (as we see them) seem to vanish when science is invoked means that there is something inherently mysterious about colour. The relationship between consciousness, our inner experiences, on the one hand, and brain processes, on the other, is somewhat similar to the relationship between yellow honeysuckle as perceived by us, on the one hand, and the physical entities and processes that are the honeysuckle, on the other hand. Just as a particular honeysuckle blossom has a perceptual or experiential aspect, and a physical aspect, so too my inner experience of seeing the honeysuckle has an experiential aspect and a physical aspect. All processes going on inside our heads – "head processes" as we may call them, to use a neutral phrase – have a neurological or physical aspect. Some of these head processes also have an experiential or mental aspect: they are our conscious inner experiences, our thoughts, feelings, desires, decisions to act, sensory experiences, imaginings – all the contents of our rich inner life.

But what precisely *is* the colour of a honeysuckle blossom? And what precisely *is* the mental or experiential aspect of the visual sensation of seeing yellow honeysuckle?

The perceived yellow of the honeysuckle is just what a normally sighted person sees in normal conditions of illumination, and nothing more. The way – and the only way – to detect the yellowness of honeysuckle (as perceived) is to be a normally sighted person yourself, and look at the honeysuckle in normal conditions of illumination. (Well, a colour photograph or film of the honeysuckle might do the job as well as the presence of honeysuckle itself.) Colour, as perceived, cannot be detected in any other way. There is no physical instrument that can detect colour as we perceive it. Physical instruments can, of course, detect *physical* aspects of colour, the capacity of an object to absorb and reflect light of various wavelengths, but that is another matter.

In a somewhat analogous way, the mental or experiential aspect of a brain process is what we detect, or become aware of, when we have that brain process occur in our own brain, and nothing more. The only way to detect the mental aspect of a particular brain process is to have that brain process occur in your own brain. There is no physical instrument that could detect the mental aspect of a brain process. The mental aspect of the visual sensation of seeing yellow, for example, is known about by all those who have normal vision: it is just what is experienced every time one sees yellow, and nothing more. What we do not at present know, of course, is precisely what the *neurological* aspects are of this sensation. The neurosciences proceed apace, much has been discovered about the neurological processes undergoing colour perception in the last few decades, and before long we may know what the neurological character of the visual sensation of yellow is. (Knowing what normally occurs in the brain when we see yellow is not the same as knowing what the sensation of yellow *is*, neurologically speaking.)

The relationship between the perceptual and physical properties of *things* (such as honeysuckle blossoms) is, then, in important respects, similar to the relationship between the mental and physical properties of *head processes*. But there are also important differences. Any number of people can see the same yellow honeysuckle blossom, but only one person can experience the mental aspect of the visual sensations I have as I look at the blossom. Only I can experience this, because only I can have this particular head process occur in my head. In order for two people to experience the same mental aspect of a head process one would have to imagine a science fiction scenario, in which Siamese twins share a part of their brains. One would have to imagine that the visual sensation of yellowness occurs in that part of the brain that is common to the two twins, and both twins experience the same mental aspect of the same head process. (And it might well be, of course, that such a neurological fantasy is not even possible in principle.)

Even though I alone can experience the mental aspects of those of my head processes that are visual sensations, we may nevertheless suppose that others can have experiences that are, in all relevant respects *the same as* my experiences. If another person is to experience what I experience when I see yellow, then that other person must have occur in his brain a brain process sufficiently similar to the one that occurs in my brain when I see yellow and *is* my visual sensation of yellow. We require, furthermore, that our two brains are, structurally and functionally, sufficiently similar, and the two brain processes in question occur in the two brains in ways that are related to the rest of the respective brain in a sufficiently similar way. Each brain process must occur in that part of the brain associated with perception, and with colour perception.

What does "sufficiently similar" mean here? The truth is that we don't at present know enough about the brain to be able to answer this question. As we grow, and grow old, our brains change; what occurs in our brains when we see yellow changes. And yet, over time, we seem to see colours in the same way (unless we become colour blind, or blind). It is reasonable to suppose that the same sensation of yellow corresponds to a vast number of brain processes different in detail: different number of neurons involved, firing in somewhat different ways. No two human brains are exactly alike – not even the brains of identical twins. But despite these neurological differences, we may, nevertheless, be able to experience exactly the same kind of sensations.

We may take the view that what matters, from the standpoint of what is experienced, is not what *stuff* a brain is made of, but rather the structural and functional features of the brain, its *control* aspects as I would prefer to say. If it is physically possible to have a brain in a body that functions like a human brain but which is made of *transistors* rather than *neurons*, then we should take the view that the transistor person has essentially similar inner experiences and states of consciousness of any human being.

It deserves to be noted that this is not the same thing as *behaviourism* – the view that a being is conscious if it behaves as if conscious. One could imagine a vast computer which has, in it, a model of a conscious brain, and is able to calculate, in real time, processes going on inside the brain – if it existed. There is also a robot with eyes and ears, in radio communication with the computer. The computer receives signals from the robot reporting on what the senses of the robot detect. The computer then, very rapidly, works out how the brain of the robot would respond – if such a brain existed – and sends radio signals to the robot which prompt the robot to act *as if in response to what the robot has perceived.* The robot behaves, we may suppose, exactly as if it is a conscious being. Nevertheless, the robot is not conscious. No head processes occur anywhere required for consciousness. In particular, consciousness is not located in the computer. The computer calculates what processes would go on inside the robot's brain if the brain existed, *but no such brain does exist*, not even inside the computer. Processes going on inside the computer involved in calculating what the brain would do, what brain processes would occur, *if the brain existed*, are very different from the brain processes themselves, if they existed. A model of a brain is not a brain. A computer that calculates how a hypothetical brain would act is not thereby exhibiting brain activity. It is thus possible, in principle, to have a robot behaving as if conscious and yet not being conscious. And that suffices to establish that the "head process theory" – as we may call the account of the relationship between consciousness and the brain that has just been expounded – is not the same thing as behaviourism.

According to this head process theory, sentient brains are unique among objects in the universe in alone having these *three* aspects: physical, perceptual, and mental. There are the physical aspects of my head processes – the neurological processes going on in my brain. There are the perceptual aspects – normally hidden from view, but which would become visible if my skull were to be cracked open. And there are the mental aspects of some of the head processes going on inside my head – aspects I alone experience as I see, feel, imagine, think, become aware.

We know that head processes going on in human brains are sentient and conscious. We suspect that mammals are sentient – and chimpanzees (and perhaps other apes) may be all-but conscious in addition to being sentient, like us. Insects, on the other hand, are almost certainly not sentient – not able to experience even rudimentary sensations, visual, tactile, olfactory, or of pain or pleasure. An important, open question, difficult to answer, is just how complex and sophisticated a brain needs to be to have sentience associated with it. In chapter eight I suggest that sentience may arise when animals are led to act as they do in response to feelings and desires of a rather general character – feelings of fear or hunger, for example, which prompt, in general terms, the kind of action that needs to be done but do not specify a precise sequence of actions of the kind performed by a spider when it builds a web. Tigers, I suspect, are prompted to hunt by hunger, but hunger does not prompt the spider to build her web. If this conjecture is correct, then sentience is to be associated with brains sufficiently sophisticated to guide their owners to act by means of rather general feelings and desires – feelings and desires that leave a great deal of scope for learning. As Tolstoy once suggested, Descartes' "I think, therefore I am" needs to be revised to read "I desire, therefore I am".[16]

This concludes my first stab at solving the problem of how the God-of-Cosmic-Value can exist within the God-of-Cosmic-Power.[17] It may be judged inadequate for at least three reasons.

---

[16] See Troyat (1970), p. 73.

[17] For earlier expositions of this two-aspect view see my (1966), (1968a) and (1968b) already mentioned in notes 1 and 5 above, and my (1984),

First, much more needs to be said about the nature of the God-of-Cosmic-Value. What exactly is of supreme value, actually and potentially, and what are our reasons for believing it exists?

Second, much more needs to be said about the nature of the God-of-Cosmic-Power. What exactly is it, and what are our reasons for believing it exists?

Third, in order to show the God-of-Cosmic-Value can exist embedded within the God-of-Cosmic-Power it is not enough to show that the experiential can exist embedded within the physical universe. In addition, it is essential to show that *we*, as beings of value, with some measure of free will, can exist embedded in the physical universe. But if everything occurs for purely *physical* reasons, in accordance with the predictions of the yet-to-be-discovered true physical "theory of everything", how can *we* be responsible for our actions, or even our thoughts? How can we exist at all?

Chapter four tackles the first of these question, chapter five the second, and chapter seven the third question.

---

pp. 174-181 and Ch. 10 (2nd ed. 2007, pp. 197-205 and Ch. 10). See also my (2000); and (2001), chs. 1 and 5.

CHAPTER FOUR

WHAT IS THE GOD-OF-COSMIC-VALUE? HOW DO WE KNOW IT EXISTS?

The God-of-Cosmic-Value is the experiential world – the world we see, hear, touch and are a part of. It is above all ourselves. It is associated with conscious life, or with sentient life more generally. The God-of-Cosmic-Value flourishes when we live life lovingly, with courage, joy, laughter, imagination, creativity, generosity and integrity, cherishing and enjoying what deserves to be (or should be) cherished and enjoyed, clear sighted about the world around us and the world within us. We are the miracle upon miracle, the holy of holies, the supreme mystery and majesty in existence, each one of us more precious by far than any work of art, cathedral, or aspect of nature. We are, as far as we know, the source of all value; everything else of value in existence is of value because of our presence in the world.[1] What is of supreme value in existence is not far away, beyond the stars, far into the past or the future, remote and inaccessible. It is, for each one of us, here and now, in the particular circumstances of our life. It is what gives pleasure, joy, happiness. Sunlight slanting through trees, a child skipping along a pavement, a gesture of friendship gratefully received, a joke, a great work of art, bird song, a worthwhile project completed: any of these might be a part of the God-of-Cosmic-Value.

"Yes, yes, yes" may be the response to this, "that is all very well, but it raises more questions than it answers. What exactly is the God-of-Cosmic-Value? What exactly is of supreme value in existence, actually and potentially? How do we choose between rival views about what is of supreme value? How do we know it really exists? How can life have any value given that it all ends in death? Are all individuals of equal value? How can justice be done

---

[1] Some of what is of value is associated with other sentient animals.

to conflicting values? Are not all value judgements irredeemably subjective, there being no such thing as what is of value *objectively*, or *in reality*? Does it even make sense to talk of what is of value, dissociated from what individual persons value? How can we make discoveries in this domain of value, learn about what is of value? Is there a realm of higher, spiritual values, glimpsed by some in rare moments of mystical inspiration, and experienced by seers and mystics?   How can we go about enhancing our capacity to realize what is of value, actually and potentially, in our life, as we live, whatever exactly may be of value, and however it may be conceived?"

In what follows I do what I can to answer these questions. Our concern is with what is of value *intrinsically*, for its own sake as it were, and not just of value as a *means* to something else of value. We may hold a medicine to be of value, not in itself, but because it is a means to health. Health in turn might be regarded as a means to other things in life of intrinsic value. Or it might perhaps be regarded as itself of intrinsic value. (Some things may, of course, be of value in both ways.)

What Exactly is the God-of-Cosmic-Value?

I give my answer to this question in the form of fourteen conjectures, each adding to what has gone before, the sum total thus becoming more substantial and specific, and therefore more doubtful, as we proceed.

(1) Everything of value in existence (i.e. the God-of-Cosmic-Value) is associated with sentient life – more particularly, with conscious life and, for us especially, with human life.

(2) It is above all individual conscious persons who are of value. What is of most value in one person's life is inherent in the rich pattern of particularities of the person's life, the extraordinarily intricate pattern of environment, deeds, perceptions, feelings, thoughts, desires, imaginings, relationships with others. The greatest poets, novelists and dramatists can only hint at the rich diversity of value inherent in a person's life. In order to come to see and to understand something of what is of value in another person's life we need the empathetic, imaginative and creative resources of a great artist so that we may enter into the person's world and, in

imagination, see, feel, experience, desire, fear, love and suffer what he or she does. We need to acquire deep personal understanding of the other. We need to be an intimate friend at least. A casual perception, a fleeting thought or feeling, of any person in life has a beauty and profundity greater by far than that of even the greatest works of art.

(3) Persons, and what is of value about them, cannot be dissociated, however, from their context and environment. Things in the environment – works of art, buildings and other artefacts, aspects of the natural world, institutions and social arrangements, may also be of (intrinsic) value – their value arising, however, as a result of their relationship with conscious persons (or at least sentient animals). Value features are not physical, but nevertheless exist objectively, as an aspect of the real world.

(4) In order to discover what is of value, we need to attend to our desires and feelings. But this does not mean that value features of things are irredeemably subjective, and do not exist objectively, in the real world. In this respect, value features are like perceptual features such as colours, sounds and tactile properties of things. In order to perceive these features, we need to experience certain sorts of sensations – visual, auditory, tactile. But that does not make these perceptual features purely subjective. As we saw in chapter 3, it is entirely reasonable to hold that colours, sounds and tactile properties really do exist out there in the world around us even though you need to have special sorts of sensations in order to perceive them. So it is with value features – and disvalue features, as we may call them – features such as kindness, brutality, generosity, grace, ugliness, tenderness, cruelty. It is entirely reasonable to hold that features such as these really do exist out there in the world around us, associated with people, actions and even things, even though you need to have special sorts of motivational and emotional responses to things to perceive them.

But not everything that we desire is desirable, and not everything that feels good is good. In order to discover what is genuinely of value we need to be prepared to subject our instinctive motivational and emotional responses to things to critical scrutiny, in an attempt to ensure that we respond to what actually exists, and not merely to value illusions and hallucinations. In this respect, again, perception of value is analogous to perception of colours and sounds. In both cases we can suffer from illusions and hallucinations, and need to subject our experiences to critical scrutiny in an attempt to detect and guard against them. The big difference is that illusions and hallucinations are far more widespread and undetected in the realm of value than in the realm of perceptual qualities.

Not only may we hold things to be of value when they are not. We may also fail to see what is genuinely of value. A piece of music may be objectively beautiful even though no one happens to experience its beauty. A human action may be objectively noble or cruel even though no one happens to experience or perceive the action in this way – possibly not even the person who performs the action. A person dies. Something infinitely precious has ceased to exist. Almost certainly, however, no one is aware of the full significance of the person's life. Even an intimate friend, a lover, can only know of aspects of the value of the person's life. Even the person herself probably failed to appreciate adequately her own value. The full significance and value of the life is something that eludes us all: and yet it is something that did objectively exist in the world, in the realm of actuality.

Just as we have to learn to see aspects of the world around us – stones, people, trees, sky – so, likewise, we have to learn to see meaning and value in the world around us, in our environment, in events, in human actions and lives. (The question of how we can best learn to realize what is of value in life will be taken up again in much more detail in chapter six.)

(5) That which is of value, for each one of us, arises in the particular circumstances of our lives. Generalities about the nature of value, such as the present ones, fail to capture the essence of what is of value just because value is specific and particular to this person, this couple, these friends, and evaporates at the level of generalities.

(6) That which is of value is incredibly diverse in character. But this does not mean that any one person's view about what is of value is as good as any other person's. Diversity of value does not imply subjectivism of value, or relativism (a point to be discussed in more detail below).

(7) Each one of us can hope to know, to be aware of, only a minute fragment of all that is of value in existence. As I have already remarked, in order to know and appreciate the value of another person we need all our powers of empathy, intelligence, imagination, perception. We need to know the other person from within, as it were, so that we have an imaginative experience of the other person's hopes and fears, joys and sufferings, relationships, struggles, feelings and desires, their life, history and world. We can only hope to have such an intimate understanding of a very few others. Most of the people we encounter are known to us only in terms of external appearances: of the inner life, where the value of the person primarily resides, we may know next to nothing. And the vast majority of the six billion or so people alive today are, of course, for each one of us, complete strangers. We can have experience of only a fragment of all that is of value, and we may even be blind to this extreme limitation of our capacity to know what is of value in existence. Our world is far, far more richly and diversely charged with value than we tend ordinarily to appreciate.

(8) Furthermore, our capacity to achieve what is of value is inevitably limited. Some suffering, failure, injustice is intrinsic to life and cannot be avoided. However fortunate and wise we may be, we will inevitably encounter limitations, failure and misfortune. And there will always be those less fortunate or wise than ourselves. The tragic dimension to life is unavoidable.

(9) Our responsibility for what is of value is also limited. Much that is of value has come into existence unforeseen and unintended. We are not exclusively responsible for all that is of value. We are not even wholly responsible for what is of value in ourselves. Even when we consciously create something of value, we do so only in so far as Nature, that which is not us, conspires with us to bring about what we intend. The creation and development of human life – the supreme source of value – is almost entirely out of our hands. Our continuing existence, our simplest deeds and thoughts, require the cooperation of Nature in a multitude of ways of which we are ordinarily quite unaware, and even do not understand (in that we are unaware of, and do not understand, the workings of our brains). There is even a sense in which Nature is wholly responsible for all that is of value

88

in existence; when we create something of value, our actions are also natural processes since we are a part of Nature and thus, in a sense, it is Nature that produces what we produce.

(10) The ability to experience, participate in and help create what is of value does not arise abruptly, inexplicably, out of nothing; rather it evolves gradually in time. Sudden flowerings of value owe their existence to long periods of prior germination and growth. We owe our present ability to participate in what is of value to the actions and efforts of millions of people who have gone before us. Almost everything of value is inherited from the past. Creation is the modification of what already exists. Our present ability to speak, to think, to be conscious and self-aware – our humanity, our self identity as persons – is, as it were, acquired from others: these things develop for us because they have already developed for others. Our existence today depends on a long process of past social and cultural evolution – and on a long process of natural evolution as a result of random variation and natural selection during millions upon millions of years. It is above all the consideration that we are a part of Nature which compels us to recognize that what is of value evolves gradually in time: abrupt creation of value out of nothing would be inexplicable, a miracle, a violation of natural law.

In seeking to discover and achieve what is of value, our task then is to develop that which is of value which already exists and has been inherited from the past. All attempts to create what is of value by means of abrupt revolutions or conversions which wholly repudiate the past are doomed to failure.

At first sight unprecedented, revolutionary achievements in the arts and sciences – achievements such as those of Shakespeare, Mozart, Beethoven, Newton or Einstein – may seem to tell against the point just made. Closer examination reveals that this is not the case. Shakespeare's plays required the prior existence of Elizabethan society, culture and language, and an already developing tradition of poetry and theatre. And most of Shakespeare's plays are based on traditional or historical themes, and modify pre-existing literature. Mozart and Beethoven both required for their work pre-existing musical traditions. Newton himself correctly declared: 'If I have seen further than others, it is because I have stood on the shoulders of giants'. Newton achieved a grand synthesis of the work of

Kepler, Galileo, Descartes and many others. Einstein's great contributions to science not only presuppose the whole framework of classical physics, the product of cooperative labour of many people over centuries but also his contributions owe much of their importance to the fact that they resolve problems buried deep in traditional classical physics and mathematics.

(11) The opposite poles of value are life and love on the one hand, suffering and death on the other hand. The supreme good in existence is living life lovingly, actively loving that which is lovable in existence and being loved; the supreme evils are suffering and death. Everything else of value in existence is organized around these two poles of good and evil.

We can help our capacity to live life lovingly to grow, or to wither and die, by what we do, what we attend to, what we strive for and value. We cannot, however, authentically command ourselves to love X, or decide to love Y at will, since real love is too dependent on spontaneous feeling and desire, out of our immediate control. We cannot therefore sensibly demand of ourselves, and of each other, that we should indiscriminately love our fellow human beings. We can however sensibly strive to create a world in which people, on the whole, treat each other, and do things together, in ways which are in accordance with certain necessary conditions for love to exist. Thus we can strive to create justice, democracy, individual freedom, tolerance, cooperative rather than hierarchical social structures, traditions of resolving conflicts based on mutual understanding, good will and cooperation rather than on bargaining, manipulation, threat or violence. In this way, love can be held to be the supreme positive value, from which all others are, as it were, derived. Justice, peace, cooperativeness, democracy, health, prosperity, enjoyment, knowledge and understanding, reason, creativity, skill, imagination, courage, beauty, sensitivity, compassion, cherishing, active concern for one's own welfare and for the welfare of others, generosity, friendliness, freedom, passion, life itself: these are all of value in so far as they are necessary conditions for the supreme thing, love.

But in addition we may hold that suffering and death are evils in their own right, as it were, and not evil only in so far as they negate the possibility of love. We do not need to appeal to the value of love in order to provide a rationale for striving to avoid unnecessary suffering and death: these endeavours carry with them their own rationale. Attempts to cure and prevent disease, to end war, totalitarianism, torture, exploitation, poverty require no further *raison d'etre* than that of bringing avoidable suffering and death to an end.

(12) No prophet, religion, revelation, book, tradition or institution is an infallible guide to what is of value – any more than our feelings and desires are such guides. Our attitudes to traditional judgements concerning what is of value ought to be analogous to attitudes to traditional scientific judgements concerning truth encapsulated in our best scientific theories: these traditional judgements, even if the best we have, nevertheless are no more than fallible, imperfect conjectures, always open to development and improvement.

(13) In addition – it almost goes without saying – there are no infallible methods or recipes for the achievement or creation of that which is of value. We cannot infallibly achieve value or know we have achieved it – even when the achievement has actually been made.

(14) The inevitability of doubt about the meaning and value of our lives ought not to be the cause of despair – any more than the inevitability of doubt in science ought to be the cause of scientific despair. Acknowledging calmly the inevitability of doubt about the meaning and value of our lives makes learning and growth possible, just as in science. Repudiation of doubt, out of fear, obstructs learning and growth. We should not seek to *rebut* scepticism about value: rather we should seek to *exploit* it in an endeavour to help increase value. As in science, so in life: we need to be so unrestrictedly sceptical, in our endeavour to realize what is genuinely of value, that we become sceptical even of the capacity of unlimited scepticism to promote the realization of value. As in science, so in life: total scepticism is to be rejected on pragmatic grounds; it cannot help. We are rationally entitled to assume that our lives here on earth are genuinely meaningful and of value, even though this cannot be verified or proved, just as we are rationally entitled to assume that the universe is, in some way, comprehensible even though this cannot be verified or proved.

How do we choose between rival views about what is of supreme value?

All too often it is assumed that judgements about what is of value must be based on some bedrock of authority or validation – the pronouncements of a prophet, the contents of a book, human nature, evolution, tradition, society, an institution (such as the Catholic Church), a body of experts (priests, gurus, artists, saints), experience, reason, intuition, inspiration, even, perhaps, moral philosophy. This whole approach is a mistake. As I have already indicated, we should treat all judgements about what is of value as *conjectures*, all too likely to be more or less inadequate, and in need of improvement. We should begin with what we value, here and now, and in trying to improve our ideas about what is of value we should (a) consult our desires, feelings and experiences, (b) consider the ideas of others, (c) consider what seem to be the best ideas in our culture and traditions, (d) learn from what transpires when we, and others, try to put ideas about what is of value into practice in life, and (e) subject all this to critical scrutiny. In particular, we should subject our aims, ideals and values to critical scrutiny when they seem problematic – when they seem to conflict with other aims, ideals or values, or when they seem to be unrealizable. We should ask why we have the aims and ideals that we do have, both in the rationalistic sense of what further aims or ideals they are for, and in the historical sense of how we came to have them in the first place. Above all, we should be especially critical of those groups, traditions and movements, religious or otherwise, that claim to know, with certainty and authority, what is of value in life – especially if doubt is held to be a sin.

I have done my best to arrive at the conjectures, expressed above, as to what is of value by myself putting something like (a) to (e) into practice. But the fourteen points above are all, of course, fallible conjectures, likely to be more or less defective in various ways, and standing in need of improvement. What criticisms are most likely to be levelled against my fourteen points?

One kind of criticism I shall not consider: the above is to be rejected because it ignores that what is of supreme value stems from God. This was dealt with in chapter one.

The claim that what is of supreme value is associated with loving and being loved is likely to be criticized on a number of grounds.

92

It may be argued, first, that love cannot be the supreme value from which all others stem. What about all those valuable activities we engage in that do not amount to expressions of love – commercial, professional, recreational? For example, how can soldiers, defending civilians from being killed, be regarded as acting lovingly as they shoot at those trying to kill the civilians? How can a judge, condemning a criminal to a long prison sentence, be regarded as acting lovingly? How can shopkeepers, bus drivers, lawyers, builders, pursuing their necessary trades or professions, be regarded as acting lovingly? Are there not countless situations in life which require us to act in ways which cannot conceivably be regarded as loving? And what about all those for whom other things are more important in life than love, such as: career, adventure, fame, wealth, pleasure, achievement in science, art, or sport, sex, comfort, security, power, the party, the movement, the nation, or pursuit of a craft or some passionate interest? Are all these people simply wrong? How can placing love on the pinnacle of value do justice to the immense variety of what is of value in life?

I tried to indicate, above, how I would respond to these criticisms. Loving, at its supreme best, involves intimacy, mutuality, a passionate sexual relationship, the giving and receiving of pleasure and joy, and care for the other. In all sorts of circumstances in life, love in this full-blooded sense is impossible, or inappropriate, even appallingly inappropriate. However, certain essential ingredients of this full-blooded sense of love are not inappropriate: some are never inappropriate. I have in mind such ingredients as concern for the welfare of others, justice, friendliness, freedom, cooperative rationality, honesty. Love can be thought of as being multi-layered: some layers, such as rationality or justice, are applicable in all situations, and ought never to be jettisoned. Others, such as intimacy, or a passionate sexual relationship, arise only in rather special circumstances, when love in the full-blooded sense is possible. Elements of love are possible and desirable in all our dealings with our fellow human beings but full-blooded love may only be possible, if we are fortunate, with one other person. On this view, to act morally is to act lovingly *to the extent that the circumstances permit*. I must add that living life lovingly does not, in my view, in general, involve sacrificing oneself for others. As I emphasized in chapter two, we should above all care for ourselves, our own welfare (as long as this does not involve trampling unjustly over the welfare of others). Loving does not require self-sacrifice; but it does not prohibit it either. A parent is not unloving in sacrificing a career to look after children.

What about the objection that placing love on the pinnacle of value cannot do justice to the immense variety of what people hold to be of value, or of what is genuinely of value in life? There are two very different objections here. Not doing justice to the immense variety of what people *hold* to be of value can only be a virtue (unless, of course, by some miracle, no one is mistaken in what they hold to be of supreme value in life). The objection that love cannot do justice to the rich variety of what really is of value is more serious. My reply is that living life lovingly has to be understood to include a wide variety of pursuits and ways of life. A craftsman can be acting lovingly in the way he pursues his craft. A scientist or scholar can pursue science or scholarship lovingly. Even a politician, perhaps, might pursue politics lovingly, in so far as the welfare of his or her constituents – and of humanity – is a primary concern. Basic ingredients of acting lovingly, in a wide variety of contexts, are caring for and enjoying what deserves to be cared for and enjoyed.

At this point it might be objected that, as a result of having been assigned this key role in the realm of value and morality, "love" has become something like a technical term, endowed with an especially rich meaning remote from the ordinary meaning. There may, perhaps, be something in this objection. If so, it is not fatal to the view I have been arguing for. If "living life lovingly" has been given an enriched meaning here, to overcome objections to the view I have been arguing for, still "loving" in this sense overlaps strongly with the ordinary meaning of the word.

Above I suggested that we should regard concern for the welfare of others, justice, friendliness, freedom, cooperative rationality, honesty as all being essential elements of loving. Notoriously, however, it is just when people are in love that they can be blind to the welfare of the beloved, careless of treating the beloved justly, far too jealous to urge freedom upon the person loved, and so swept up by passion to become, in some circumstances, unfriendly, uncooperative, irrational and dishonest. And yet a person who treats the beloved in all these ways might still be regarded, by many, as being in love, or loving the beloved. Here, then, are a number of ways in which the notion of "loving" I have been appealing to here may differ dramatically from what is ordinarily meant by "loving". None of this discredits the notion I have been appealing to, or the view I have been arguing for. We should conclude, rather, that some popular notions of "love" are defective in various ways.

With these qualifications and reservations understood, it becomes possible to declare that a good world, a civilized world, is a *loving* world. A loving world, in this sense, is not one in which everyone loves everyone else. It is rather one in which most individuals have loving relationships, the capacity to live life lovingly is widely distributed, there is a culture of loving, and in the public domain the appropriate essential ingredients of loving hold sway – ingredients such as individual freedom, justice, and cooperative rationality.

What is to be said about attempts to put this philosophy of loving into practice in life? There is perhaps far more loving going on in the world than we may realize. What gets attention in the media is not loving, or the products of loving (unless it is scandalous), but the opposite: acts of violence, crime, war, natural disasters, corruption, the heartless aspects of the modern world. (I speak here of media in democracies: media in dictatorial nations tend to heap praise upon the Leader, his administration, and the brave forward march of the nation.) We take the view, perhaps, that living life lovingly, as best we can, is the normal state of affairs; it is the gross departures from this normal state of affairs that needs reporting. Be that as it may, I have here just three remarks to make about living a life of loving.

First, it is probably correct to say that those who live life lovingly most successfully do so instinctively, from the heart, and not because they have decided, intellectually, that this is the best philosophy of life to put into practice. I don't mean successful loving is all a matter of instinct, and does not involve thought, care, perceptive attention to others. On the contrary, these seem to me to be essential to loving. What I mean, rather, is that those who live lovingly are most likely to do so, not because of an intellectual decision, but because of temperament, upbringing, fortunate circumstances conducive to such a way of life. (But all such sociological theses are wildly conjectural, at best only true on average, there being endless exceptions.)

Second, a peculiar danger awaits those of us who consciously seek to put this loving philosophy into practice in our life. The danger is hypocrisy. If we pride ourselves on being loving, so that much of our sense of self-worth is bound up with our idea of ourselves as being loving, we may be especially reluctant to acknowledge to ourselves those occasions when we have acted in distinctly unloving ways. This is a general point. Whatever we pride ourselves on being – good, wise, selfless, charismatic, etc. – there will be a tendency to blind ourselves to those occasions when we are nothing of the kind.

This said, it would be wrong to conclude that adopting a philosophy of loving, and making the intellectual decision to live it, can have no positive effects in one's life. We can decide to modify our aims and actions, and choose circumstances, that tend to be conducive to living a loving life.

Third, the thesis of this book is, of course, that our fundamental task in life is to live life lovingly. This includes contributing, with others, to the great world-wide task of making progress towards a loving world. This, as I see it, is the religious task of helping the God-of-Cosmic-Value to flourish within the God-of-Cosmic-Power. A glance at the world stage indicates that we have a long way to go. It is above all here, in the public domain of politics, business, international relations, agriculture and industry, that adoption and implementation of a philosophy of loving might make some difference – as I shall argue in chapters six and nine.

These, then, are the kind of considerations that can be brought forward in support of, or in criticism of, a particular conjecture as to what is of supreme value in existence. I do not suppose that the above sketchy remarks will convince many of the correctness of my view (unless, perhaps, already convinced). But let us, for the sake of what follows, take this as a token view, an exemplar of what a conjecture about the nature of the God-of-Cosmic-Value may be taken to be. What matters of course is what really is of value, actually and potentially, in existence. Conjectures about the matter deserve to be taken that more seriously as they come to reflect the reality of what is of value that more accurately and faithfully.

How do we know that that which is of value really exists?

If you are happy, if you are engaged in what seem to be worthwhile and enriching enterprises, and if you are among friends and loved ones, it may well seem absurd to raise questions about the value of life. What could be more blatant, more solidly known and experienced, more an integral part of one's existence, the very stuff of one's life, than the value of what one is a part of? To call this into question, and doubt its existence, may seem ludicrous, even faintly sacrilegious.

But there are, for most of us perhaps, other occasions, other moods, when such doubts press upon us, and we are indeed haunted by the feeling that it is all, ultimately, pointless and meaningless. Some people live with black despair for years. One's own life can seem a hollow mockery, and the rich, rewarding life of others can seem fraudulent. The whole vast panoply of the world can seem no more than a vast, frenzied distraction, an attempt to disguise the emptiness of it all. There is no value in the world, only at best the successful creation of the illusion of value. Ultimately there are just cold hard physical facts, things happening one after the other, all amounting to no more than a tale told by an idiot.

How can this dreadful possibility be decisively *refuted*? I am not sure that it can. All our knowledge, ultimately, is conjectural in character, as Karl Popper tirelessly argued, including any knowledge we may have about the existence of what is of value. As I tried to indicate in point (14) above, the best we can do from an intellectual standpoint, it may well be, is to grasp at the straw of a pragmatic argument. We should not attempt to *refute* scepticism about value but, on the contrary, should give it free reign to help us unmask fraudulent claims to value, so that we may learn and improve the value of our lives. The really valuable life is not the one confidently immune to doubt about its worth; it is rather, at least, the life that is always open to the possibility that what seems to be of value is actually nothing of the kind – and thus a life which is open to the possibility of learning what really is of value. If we are to live a life of value, or if we are to give ourselves the best chances of living such a life (in so far as such a thing is to be had), we need calmly to sustain doubt about the value of our lives, be open to the possibility that it is not entirely (or perhaps not at all) what it seems, and not let awareness of such possibilities be discouraging. The crucial intellectual point, both in the realm of value and in the realm of knowledge about fact (even in science) is that scepticism is not an enemy to be defeated. It is rather the friend which makes learning and progress possible – a point that goes all the way back to Socrates. To this one needs to add that it is important not to be *relentlessly* sceptical, *unsceptically* sceptical as one might say. If we are relentlessly sceptical about everything, learning and progress become impossible. Scepticism is rational in so far as it facilitates learning; it tips into irrationality and neurosis when it negates the very possibility of learning by doubting *everything*. To be open to the possibility that some specific thing which seems charged with value is actually devoid of it is rational. To doubt seriously that anything at all is of value is irrational because it destroys the rational function of scepticism, which is to help us to learn, to make progress, to enhance the value of our lives. We need to be, not just sceptical, but sceptical of scepticism itself – so that we appreciate that doubt deserves to be dismissed when it is such that it cannot help us learn or make progress.

This argument does not prove that life is of value. Rather, it provides grounds for holding that it is rational to uphold the conjecture that life is of value, in the teeth, even, of moods of the blackest despair (poor comfort, I admit, for those in the grip of despair).

How can life have any value given that it all ends in death?

We live for a few decades if we are fortunate, and then we die, and everything that we are, do, have, think, feel, comes to nothing. There it sits, up ahead in our life, annihilation, the ultimate personal catastrophe, and we can do nothing to escape. It sits there, grinning at us, making our life a mockery. What possible value can life have if everything we are and do ends in nothing?

We may have children, who may survive us, but they too will die, and their children in turn. We may live on for a while after death in the memories of those who knew us, but not for very long. We may make contributions to science, to art, to business, society or sport, which live on after our death, but this is the preserve of the few, for most of these few memories of the person does not last long, and in any case, even for the rare Shakespeare, Mozart or Einstein, this form of immortality hardly seems to compensate for personal death. As Woody Allen remarked "I want to be immortal in my life, not through my films".

And in any case eventually, sooner or later, the human race and everything it has achieved, done and suffered will come to an end. We cannot draw sustenance from the value of the lives of others, from our culture and way of life persisting long after our own death, for this too will one day die, and come to nothing.

What value, then, can life possibly have if ultimately it all ends in nothing?

The answer is easy to say, not so easy, always, to live. What is of value in a life is to be found in the life itself, not in some external, eternal end-product. It is a common but profound error to hold that only that which is eternal has real value, or the value of something is to be measured in its longevity. Put into the cosmic scheme of things, human life is extremely brief, but that does not deprive it of value. Ideally, we should live our life calmly aware of the inevitability of death, and should not allow ourselves to be provoked into a frenzy of distraction or denial. Ideally, we should be able to feel that the value of what exists here and now is such that its blaze outshines black eternity, and we can look at nothingness in the face and not mind its eventual inevitable arrival.

The real problem, perhaps, is not so much fear of death as fear of not really living, and then dying. Living and then dying is not as bad as never really living at all. We fear death because it reminds us that unless we make a start on our real life soon it will be too late: we will die before we have begun to live. The time of death sounds the knell on all those glimpsed but never realized amazing possibilities.

It is not hard to understand why those influenced by modern western culture should find it especially hard to come to terms calmly with death. This is hard enough as it is, whoever you may be, and whatever your cultural background. Death *is*, near enough, the ultimate personal catastrophe. Darwinian evolution has planted in us a fierce desire to survive. The discovery in human pre-history that this desire must ultimately be defeated, and there is no escape, must have been traumatic indeed. The archaeological records indicate that, from the earliest beginnings of culture, attempts were made to deny the reality of death by imagining a life after death. Most religions have some kind of account of life after death – and this is true, of course, of Christianity as well. Christianity has, of course, exercised an immense influence over the evolution of western culture. In particular, that aspect of Christianity, which emphasizes that the whole point of life lies in what happens afterwards, has been enormously influential. It has infused into western culture the attitude that the moral life involves foregoing immediate gratification for distant goals. This is behind the Protestant work ethic, behind the drive to devote a life to amassing a fortune, building up a business empire, establishing a reputation in a profession, leaving behind a body of great art or science. If the basic tenets of Christianity were true, this attitude just might be the correct one to adopt. But with the Christian God cut in half, and prospects of life after death no more plausible than prospects of Father Christmas paying one a visit down the chimney, the attitude of foregoing gratification until some distant time in the future does not make much sense. Life is too short and precious to be put on hold. We have to live now. If we do not we may well come to fear death in an especially severe form, because death demonstrates the disaster of failing to live now, while we still can. The enormous influence of Christianity on western culture means, in short, that those of us influenced by this culture are likely to be especially prone to the disease of foregoing life today for distant goals and, granted we do not believe in life after death, especially prone to the severest forms of fear of death, and peculiarly ill-equipped to come to terms calmly with the inevitability of death. We will experience the thought of death as an acid that insidiously eats away all meaning and value from our life and leaves nothing but an empty shell. It will seem that death annihilates value.

An interesting confirmation of this is that many Christians do indeed argue that if God does not exist then life must be meaningless. The value of life, according to this attitude, lies not in life itself, but elsewhere, in God, and in life after death. Cut God in two and cancel life after death, and all value in life will seem, inevitably, to drain away.

I do not mean the above to imply, incidentally, that we should live only in the present and forego all long-term plans, goals and pursuits. That would simply be the flip side of the coin. Quite to the contrary, it seems to me that the worthwhile life is likely to involve some long-term goals and pursuits. The vital point is that, in engaging in such pursuits we should continue to live now, and not forego life now for some distant dream time in the future. The mistake is to think our life now hinges on some future occurrence or state of affairs.

We have been considering whether death cancels meaning and value, and I have answered: No, it does not, although it may well feel as if it does for those of us who live in a Christian-influenced world, for thoroughly understandable reasons. Quite different, of course, is fear of death when we are in danger, or fear for others in danger. On these occasions, when death is on hand as an immediate possibility, and is not merely an eventual inevitability, we are likely to find we are brought face to face with how infinitely precious life is, and thus with just how much we have to lose.

Additional reasons can be given as to why fear of death – or fear of only partly living before dying – may have intensified as history has unfolded. In pre-historical times we lived in small scattered hunting and gathering tribes. Everyone in the tribe would have known everyone else, everyone would have taken for granted the same way of life, the same view of the world, customs and values. No particular reason exists to fear one has made the wrong choices in life as only one kind of life is available. Life is what everyone lives. Then history occurs. Tribes coalesce, cities and nations come into existence, modern methods of travel and communication develop, and each one of us, potentially, becomes a citizen of the world. We are aware, to a greater or lesser extent, of a myriad of different cultures, ways of life, occupations, values, fortunes and misfortunes. We are aware, too, of rapid change: life was different fifty years ago, and will be different again in fifty years time. Everything may seem arbitrary, temporary, contingent, no more than brief current fashion. A feast of choices lies before us, some – known via TV and the movies – being alluringly glamorous, drenched in fame, wealth and power. Our actual capacity to choose may, however, be strictly limited: we are born in the wrong place, to the wrong parents, at the wrong time, with the wrong talents and opportunities. Our situation is profoundly different from that of the hunter and gatherer. Some of us today have wealth and opportunities beyond the hunter and gatherer's wildest dreams (although hundreds of millions do not).

But even if we do, these riches and opportunities may seem to turn to ashes because what seem to be the real prizes are beyond our reach. We have to choose, but choice itself is fraught with uncertainty, contradictory guidelines providing contradictory advice. Our situation is highly conducive to our coming to feel we have failed, we have been condemned to a shadow of what our life could have been.[2] Or, if we are young and such life choices lie in the future for us, we may well feel that our chances of having a really worthwhile life are slim indeed. What is the appropriate response: frantic effort, or despair and distraction? In any case, our modern circumstances are such that many people are likely to feel they are only partly living: death will seal forever those alluring unlived possibilities. Death shouts not "You die!" but the much more terrifying "You never lived!".

Are all individuals of equal value?

People are not of equal value. How could we hold an Albert Einstein, a Wolfgang Mozart or a Nelson Mandela to be of value equal to monsters such as Adolf Hitler or Stalin? We are not even born equal. Biology does not respect justice. What can be said, of course, on behalf of equality is that we ought all to be equal before the law, and ought all to have equal civil rights. I would say, too, that a good society is one in which there is far greater equality of wealth and power than in the world today. Enforcing equality too rigorously inevitably undermines freedom, and even equality itself in that a special class is required, with privileges above the rest, to enforce the equality of the rest, as in the old Soviet Union. Some inequality is the price that must be paid for liberty, but that argument does not justify the extremes of inequality we see in the world today.

---

[2] Karl Popper has argued that the breakdown of tribal society, and the beginning of the "open society" – a society that tolerates diversity of views, values and ways of life, induces in people what he calls *the strain of civilization*. This is "the strain created by the effort which life in an open and partially abstract society continually demands from us – by the endeavour to be rational, to forego at least some of our emotional social needs, to look after ourselves, and to accept responsibilities". See Popper (1969), vol. 1, p. 176.

We may also hold that individuals blessed with extraordinary talents should so develop and deploy their gifts that their fruits become accessible to all, and should not exploit these gifts primarily for their own benefit, at the expense of others. We can all benefit from what exceptionally talented individuals may produce, and hence it is important that we appreciate exceptional talent as potentially in all our interests, and do not covertly try to suppress its development because of an anti-elitist attitude which holds that rare geniuses are an offence to the equal society.

How can justice be done to conflicting values?

In chapter six I will emphasize that basic aims, ideals and values are liable to be inherently problematic. This, as we shall see, has profound implications for rationality. It means reason must be such that it helps us resolve problems – clashes and conflicts – in our basic aims and ideals as we live. But can we be sure that these clashes and conflicts can always be resolved, even in principle? One value system, somewhat right wing in character, might prize individual initiative and reliance, freedom from government interference, personal achievement, prosperity and rewards to those who work hard at the expense of equality. Another somewhat more left-wing value system might prize equality, solidarity, fraternity, strong democratic government, even at the expense of liberty. How, even in principle, can a rational choice be made between these two value systems? Each will be self-validating. Each will provide its own criteria for assessing value systems in terms of which each comes out on top. How then can an objective choice be made?

The answer is to adopt the hierarchical, aim-oriented conception of rationality to be expounded in chapter six. This stipulates that, in order to resolve the conflict between these two value systems, we need to specify the large area of agreement between the two views, and then seek to assess the relative merits of the areas of disagreement against (a) the area of agreement, and (b) experience – what ensues when each view is implemented, or *lived*. In principle, as we shall see, this provides the means for resolving such conflicts, although it may not do so, of course, in practice.

Are not all value judgements subjective, there being no such thing as what is of value *objectively*, or *in reality*?

If all value judgements are irredeemably subjective[3] – so that no judgement concerning what is of value represents how things really are better than any other – then *in reality* nothing is of value. Objective reality is denuded of value. Value is a subjective illusion we invent to lull ourselves into the false belief that what is of value really does exist. The God-of-Cosmic-Value, in short, does not exist. It is an illusion.

But why should anyone hold that value judgements are subjective? For many, there are apparently decisive grounds for holding this view – or the related view of relativism. These come in three categories, which I shall call moral, metaphysical and epistemological. In what follows I expound these objections, and then show what is wrong with them.[4]

To begin with, we may hold it to be immoral to proclaim the existence of objective value, and then invoke it in an attempt to influence the conduct of others. The mother tugs the restless child's hand and exclaims "Be good!" when what she really means is: "Do what I want you to do!" The act of telling the child to be good is an act of manipulation and deceit. The same thing happens when the authorities tell the public to "cooperate with the authorities": this does not mean "work in partnership (i.e. cooperatively) with the authorities"; it means "Do what the authorities tell you to do". Moral systems can be regarded as systems of control and exploitation, put about by those in power to induce others to act in the interests of those who hold power. Interpreting such moral systems as "objective" further obscures the

---

[3] Subjectivism about values and morality is a widely held and influential view. It was the dominant view among moral philosophers in the 20th century. Subjectivist views are to be found in Ayer (1936); Stevenson (1944); Hare (1952); Nowell-Smith (1952); and Mackie (1977).
[4] When I first argued for value realism I thought I was a lone voice crying in the wilderness: see my (1976a), pp. 138-146 and 242-254 or 2nd edition (2009), pp. 140-148 and pp. 246-258; (1984) or (2007a), ch 10. I have subsequently discovered others who have argued for the view: see Bond (1983); McDowell (1998), Part II; Brink (1989). For an excellent review article see: Little (1994).

manipulation and deceit that is involved; it makes it that much more immoral.

Similarly, it may be argued, those who proclaim the existence of objective values do violence to liberalism in that, instead of questions of value being left to individuals to decide for themselves, such questions are decided by the authorities, the experts, those who are in a position to "know" what is best for the rest of us. Objectivism, or realism, it may be argued, is authoritarian, even totalitarian in spirit, a ploy used to indoctrinate and enslave. Objectivism provides a ready justification for imperialists and religious fanatics, for those who know with certainty what is right, and on that basis strive to gain power over others by means of force, persuasion or terror.

Yet again, it may be argued, at a milder level, objectivism, in the field of the arts leads straight to elitism. Those who are in a position to do so proclaim that those arts that they enjoy are objectively of greater aesthetic value than those enjoyed by others, and on that basis ensure that what they enjoy receives much more patronage and state funding.

In addition to these moral objections, there are also metaphysical objections. What are these mysterious value facts, in virtue of which value statements are either true or false? What are value properties, and how are they related to physical properties? Do we, with G.E. Moore, think of the Good as an unanalysable property which cannot be defined?[5] Or do we, even more radically, with Pirsig, think of Quality as the basic stuff of existence, indefinable, neither objective nor subjective, from which everything else emerges?[6] Are we to suppose that value is some sort of mysterious invisible fluid, valuable things being soaked in it, valueless things being bereft of it? Might chemists one day distil drops of this precious fluid in a flask? The whole idea is surely preposterous. And even if this mysterious value substance or property existed, it would remain a mystery how we can come to know that some things possess it; and even if we could know

---

[5] See Moore (1903).
[6] Pirsig (1974).

this, it would be utterly mysterious why we should especially value things that are rich in this mysterious property of value.

Finally there is the epistemological objection. If objective value exists, then it ought to be possible to determine, objectively, whether something is or is not of value. It ought to be possible to decide disputes about what is of value by an appeal to the objective value facts, much as factual disputes can be decided in science. But notoriously, disputes about what is of value are endless and seem inherently unresolvable. This, again, seems decisive grounds for rejecting objectivism.

In order to see what is wrong with these three reasons for rejecting value realism, it is essential to recognize that a number of different versions of objectivism can be distinguished; most succumb to the above moral, metaphysical or epistemological objections, but one does not.

In order to overcome the moral objections to objectivism we need to recognize that there are at least *three*, and not just two, positions, namely:

1. Dogmatic Objectivism: There are objective values, we know what they are, and anyone who disagrees must be (a) taught better, (b) converted, (c) conquered, or (d) assassinated.

2. (Dogmatic) Relativism: What is wrong with Dogmatic Objectivism is the objectivism. There are no objective values, there is only what people desire, prefer or value.

3. Conjectural Objectivism: What is wrong with Dogmatic Objectivism is the dogmatism! Precisely because values exist objectively, our knowledge of what is of value is conjectural in character. If two parties disagree about what is of value, the chances are that each has something to learn from the other.

Dogmatic objectivism is the sort of view upheld (in its milder forms) by the Victorians when confronted by primitive people. Victorians not only believed in the existence of objective values, but "knew", beyond all doubt, that the correct values were those of Victorian England. Primitive people, with very different systems of values were, in the eyes of Victorian travellers and anthropologists, simply wrong, ignorant and primitive. Today it is, typically, various sorts of religious fundamentalists who uphold versions of dogmatic objectivism.

Relativism arises as a result of a reaction against dogmatic objectivism. It seems appalling that people should be so convinced of the correctness of their views about what is of value that they feel justified in converting or conquering everyone else so that they too come to live by and believe in these views – even to the extent of feeling justified in eliminating those who refuse. People proselytize their values, their religion and way of life so aggressively because they believe they have the might of objective value behind them, in the form of gods, God, the Tribe, The Race, the chosen People or Class, the Nation, History, Civilization, or whatever. These are regarded as objectively existing embodiments of value, and it is this, so incipient Relativists believe, which leads to the drive to dominate and convert, to offend basic principles of morality and liberalism. It is the value-objectivism of dogmatic objectivism which is the cause of the problem, Relativists argue, and as a result defend value-subjectivism. The whole idea of value existing objectively, of value-judgements being objectively true and false, is a nonsense: there are simply a multiplicity of preferences of people, some embodied in diverse value-systems, no one being better or more correct than any other, in any objective sense. Those who belong to so-called "western civilisation" should regard so-called "primitive" people as merely different, not inferior.

But Relativism, despite its good intentions, is hardly an improvement over Dogmatic Objectivism. Given the latter view, it is at least possible to hold that the imperialist actions of the Victorians were objectively wrong. Given Relativism, this becomes impossible; one can only say that these actions are not to one's own personal taste. Relativism seems to defend liberalism and tolerance against imperialist aggression, but the defence destroys the very possibility of declaring liberalism and tolerance to be morally good and imperialist aggression to be morally bad. The defects of Relativism defeat its own good intentions. And there are the other adverse consequences to take into account as well: the annihilation of value, the cancellation of the possibility of learning in the realm of value. If any view about what is of value is as good as any other one, then nothing is of value objectively, in reality, and there can be no learning about what is of value (except learning that nothing, ultimately, is of value).

It is important to note that Relativism objects to the objectivism of dogmatic objectivism, and not to the dogmatism. There is indeed a sense in which the transition from dogmatic objectivism to relativism intensifies the dogmatism. A Dogmatic Objectivist is convinced that he is right and those who disagree are wrong; at the same time he holds that this is a significant issue, one worth going to war and dying for, and thus certainly not meaningless. In other words, it is definitely meaningful that he might be wrong about what is of objective value; but he knows he is right. For the Relativist, however, it is meaningless that one can be wrong about one's personal preferences: what higher authority than one's self could there be? There are of course somewhat trivial senses in which one can be wrong: one may be wrong about what one's actual preferences are; or one's actual preferences may be the result, in part, of false purely factual beliefs. Putting these points on one side, it is, according to the Relativist, meaningless to say that one person's preferences are right, another's wrong. In this respect, yet again, Relativism is hardly an improvement over Dogmatic Objectivism.

Relativism is right to object to Dogmatic Objectivism, but wrong to object to the objectivism of the view. It is the dogmatism of Objective Dogmatism that is objectionable, not the objectivism. It is the dogmatism, the absolute conviction in the correctness of one's own position, that makes it possible for one to be convinced that non-believers should be (a) taught better, (b) converted, (c) conquered, or (d) assassinated. Not only does Relativism misallocate what is wrong with Dogmatic Objectivism; it actually has the effect of intensifying what is wrong, as we have seen. Relativists may hope that general acceptance of their view would promote tolerance, but the hope is misplaced. Relativism puts those who seek to convert, conquer or assassinate on a par with those seek to live cooperatively and tolerantly with their fellow human beings. Furthermore, general acceptance of Relativism is as likely as not to sabotage growth of tolerance, since tolerance is, by and large, something that needs to be learned and, as we have seen, Relativism cancels the very idea of learning in the realm of value.

Dogmatic Objectivism and Relativism make the same blunder: both take it for granted that objectivism leads to dogmatism. In fact precisely the opposite is the case: objectivism demands that we recognize that we cannot know for certain what is, and what is not, of value; at best our value judgements must be conjectures. If there really are value features of things that really do exist whether we perceive them or not, it becomes all but inevitable that we will, more or less frequently, get things wrong. Just because the physical world really does exist, we often make mistakes about it; we do not have an infallible access to all that there is. On the contrary, much of the fallible knowledge that we do possess about the physical universe has only been won as a result of centuries of effort by science. What possible justification could there be for supposing that the situation is different as far as value features of things are concerned? If such features really do exist, then surely here too we must acknowledge that we cannot hope to be infallible, that our views about what is of value are all too likely to be more or less wrong, and hence such views need to be held as conjectures. Objectivism, in other words, all but implies conjecturalism, and demands that one rejects dogmatism.

As long as we believe that only the two views of Dogmatic Objectivism and Relativism are possible, we are forced to choose between them, even though both, as we have seen, have highly undesirable consequences. The all important point to appreciate is that a third view is available, Conjectural Objectivism, which need have none of the moral and intellectual defects of the other two views. Dogmatic Objectivism and Relativism, as we have seen, clash with or undermine liberalism. By contrast, Conjectural Objectivism, far from clashing with liberalism, may be held to be necessary for liberalism. For, granted Conjectural Objectivism, we may conjecture that it is people, and what is of value to people, that is ultimately of value in existence. In other words, the basic tenet of liberalism, which one might state as "It is individual persons that are of supreme value in existence", needs to be formulated as a conjecture about what is objectively of ultimate value, and for this one requires Conjectural Objectivism. If

109

Relativism is presupposed, the basic tenet of liberalism disintegrates into nothing more than a personal preference.[7]

In order to overcome the metaphysical objections to Objectivism we need to exploit a point made in chapter three: there are two very different ways of drawing the distinction between objective and subjective, two meanings that can be given to "objective" and "subjective". The first distinction has to do with whether something really exists, or does not exist (but only appears to exist). The second has to do with whether something is utterly impersonal, unrelated to human beings, or whether it is in some way personal, or related to human beings.[8] The all important point is that something may be subjective in the second sense, but objective in first sense. That is, something may be related to human concerns, aims or physiology and yet, at the same time, may really exist out there in the world. Value features are of this type: like perceptual qualities such as colours, they are related to human concerns and aims, but really existing for all that.

Let us call the first meanings of "objective" and "subjective", connected with existence and non-existence, "existential objectivity" and "existential subjectivity".

If some object or property is existentially objective, then it really does exist; if it is existentially subjective, then it does not really exist even though it may appear to do so, or may be thought by some to exist. Tables, trees and stars are existentially objective; ghosts, demons and spells are existentially subjective.

Let us call the second meanings of "objective" and "subjective", connected with being human-unrelated and human-related, "humanly objective" and "humanly subjective". An object or property is humanly objective if it is wholly impersonal, unrelated to human aims, interests, experiences or physiology; it is humanly subjective if it is related to human aims, interests, experiences or physiology. Physical entities and properties, such as stars and atoms, mass and electric charge, may be taken to be humanly

---

[7] For an earlier discussion of these issues see my (1984), pp. 255-258; and my (2001), ch. 2.

[8] For an earlier formulation of this distinction see my (1966), pp. 310-311.

objective, in that these objects and properties are entirely unrelated to human interests, aims or physiology. By contrast, works of art, constitutions, legal systems and languages are all humanly subjective in that these objects are all quite essentially related to human beings. Furthermore, properties such as poisonous, green, delicious and friendly are humanly subjective in that these properties are all human-related.

The crucial point in all this is that, even though something is humanly subjective this does not mean that it is existentially subjective. On the contrary, it may be existentially objective. Bach's St. Matthew's Passion, Britain's constitution, legal system and language all exist (are existentially objective) even though they are also human-related objects (i.e. humanly subjective). Arsenic really is poisonous, grass really is green, zabiogne really is delicious, and Einstein really was friendly (i.e. all these properties are existentially objective) even though these properties are human-related (i.e. humanly subjective).

It is into this category of existential objectivity and human subjectivity that value features fall. Like colours, value features really do exist out there in the world; but also like colours, value features are human-related.

If we hold that there is just one distinction between the objective and the subjective, we thereby make it impossible to declare that colours, and value-features of things, are existentially objective but humanly subjective. Declaring value-features to be objective commits us to declaring them to be human-unrelated, like mass or electric charge, which is absurd; but also, declaring value-features to be subjective commits us to declaring that they do not really exist, which seems equally absurd. The above dilemma, in short, arises as a result of failing to appreciate that there are two quite different distinctions between objective and subjective: the dilemma is readily solved once one appreciates this point, which permits one to say that value-features are objective in one sense (really existing) but subjective in another sense (human-related).

Put another way, once we recognize that there are two distinctions between objective and subjective to be made, then, in declaring values to be objective there are two possibilities. We may mean that values are existentially objective and humanly objective: let us call this view impersonal conjectural objectivism. Or we may mean that values are existentially objective but humanly subjective: let us call this view human-related conjectural objectivism. The above metaphysical objections to objectivism apply devastatingly to impersonal conjectural objectivism: it is indeed absurd to suppose that a value-fluid exists in the universe, which chemists might one day distil in a flask. But these metaphysical objections fail completely when directed against the more modest view of human-related conjectural objectivism. The value-features of things are as familiar, unmysterious and non-metaphysical as colours, sounds and smells. In order to perceive value features we may need to have emotional responses, just as in order to see colour we need appropriate visual responses: but in neither case does this mean that the property is existentially subjective – though it does mean it is humanly subjective.

Typical familiar value-features of people are: friendly, mean, jolly, stern, witty, courageous, warm-hearted, dull, frivolous, shifty, kind, spontaneous, strong-willed, earnest, gloomy, calculating, mischievous, cold, boring, gushing, loyal, ambitious, argumentative, generous. These are both descriptive and value-laden, factual and imbued with value. People, like works of art in a somewhat different way, are essentially value-imbued, morality-imbued things: we cannot describe a personality, we cannot state facts about a personality, without employing value-imbued factual terms of the kind just indicated, any more than we can describe a work of art as work of art without employing analogous aesthetic terms, value-imbued factual terms.

Those who wish to maintain the traditional distinction between fact and value will argue that terms such as the above can always be interpreted in two ways, first in a purely factual, non-evaluative way, and second in an evaluative and non-descriptive, non-factual way. We can describe without evaluating, and in adding an evaluation we do not provide additional factual information, we do something quite different, namely evaluate.

It should be noted that I have not argued for the existence of value-features; I have confined myself to rebutting arguments against the view that value-features really do exist in the world. This, in my view, is the crucial task that needs to be performed.

No one, I believe, would take relativism or subjectivism seriously if they were not persuaded that value objectivism is untenable. What needs to be done is not to prove that value features of things really do exist (a hopeless task in any case), but rather to prove that arguments against objectivism are invalid.

Continuing in this vein, let us consider what grounds there are for insisting that the above value-laden factual terms must be split into two distinct parts, the factual and the evaluative. Consider "friendly". On the face of it, this is doubly evaluative, first because friendliness may be deemed to be a desirable quality in a person, and second because friendliness may be deemed to be such that a genuinely friendly person, at the very least, acts in a moral way towards other people. One cannot be friendly and mean, friendly and cruel, at one and the same time. What obliges us to split off a purely factual, non-evaluative meaning from the evaluative, moral meaning?

Doubtless this can be done. We can, for example, render "friendly" purely factual by specifying some set of values and interpreting "friendly" in terms of this set, there being no presumption that this set embodies what is really of value.

But what grounds are there for holding that this must be done, apart from the mistaken idea that value-features of things cannot exist?

In my view, a particularly strong reason for holding that value-features exist, for supporting human-related conjectural objectivism, arises from the following sort of consideration (already indicated in (4) above). Think of a friend or relative you have known personally, neither a saint nor a fiend, who has lived her life, and has died. A number of people have known this person, in different contexts, and to differing degrees. The deceased person will have revealed different aspects of her personality to these lovers, friends and acquaintances. No one, it is all too likely, knows all that there is to be known about this person. No one knows all the good qualities of this person. Even the dead person, when alive, may not have been aware of her good qualities; she may have undervalued herself, been too aware of failings and insufficiently aware of countless acts that have brought pleasure, delight or happiness to others. No one sees all that is of value in

113

this person. But we should not conclude that it therefore does not exist. To do so would have the dreadful consequence that it is only those who are widely believed to be of value who really are of value, and those who have quietly contributed much to the quality of people's lives, unnoticed and unsung, are nothing, and have done nothing.

In the realm of value, to believe that to be is to be perceived, which is what subjectivism and relativism amount to, is to be a cynic and nihilist of dreadful proportions. Early 21st century life suffers horribly from these doctrines. Even fanatical fundamentalism may be seen as a sort of hysterical reaction to the cynicism and nihilism implicit in value subjectivism and relativism, widely upheld because philosophical blunders (indicated above) appear to leave liberalism, and a sane scientific outlook, no alternative.

So much for my criticisms of the moral and metaphysical objections to value realism. I turn now to a consideration of the third, epistemological objection.

The epistemological objection, stated above, is that if value features of things really exist then it ought to be possible for people to agree as to what they are. Notoriously, people disagree, and there appears to be no procedure for achieving agreement, as in science or mathematics. Hence objective values do not exist.

The lack of universal values is often taken as a strong argument for Relativism, and Objectivists often assume that, in order to establish their position they must demonstrate, somehow, that there is some set of values that arise universally in all cultures. But all this is a mistake.

The physical universe exists independently of us; here, unquestionably, there are objective facts.[9] But when it comes to

---

[9] Some philosophers and historians of science have questioned whether there are objective facts about the physical universe. I assume here that this "cognitive" version of Relativism is untenable. If based on the anthropological evidence that different societies have held different views about the universe, cognitive Relativism tends to become inconsistent in that it recognizes the existence of people, presumably living on the surface of the earth, these being objective facts, and then goes on to assert that there are no objective facts. If this is put forward

cosmological theories concerning the nature of the universe, we do not find that there is some universal theory, accepted by people in all cultures at all times. On the contrary, we find an incredible diversity of views. But this does not mean that there is no such thing as the true nature of the universe; it just means that this truth is inaccessible, difficult to get hold of (and hence the need for science). Science by no means puts an end to diversity of views. In science, too, one finds a number of conflicting cosmological views.[10]

The same point arises in connection with value-features of things. Long-standing, widespread disagreement about what is of value does not mean that there is no such thing as that which is of value objectively; it just means that it is more or less inaccessible, more or less difficult to determine or establish.

To this it may be objected that there is still a big difference between the two cases. As far as the physical universe is concerned, different societies and cultures may have produced radically different cosmological theories; and even different physicists may defend different theories: nevertheless in this domain we possess the means for resolving debates between conflicting views. In gradually improving knowledge, science sooner or later decides between diverse conflicting hypotheses.

But in the realm of value, nothing of the kind is discernable. Notoriously, different people, different societies and cultures disagree radically about questions of value, and no amount of argument or experience seems capable of resolving these conflicting views. There is no science of value; the very idea seems somehow absurd. Do not these considerations support the view that in the realm of value we are concerned merely with various purely subjective tastes or desires, there being no such thing as an objectively existing value feature?

---

as a *reductio ad absurdum* argument, the assumption that people exist on earth leading to the conclusion that there are no objective facts (because people disagree as to what these facts are), the straightforward reply is that disagreement about facts does not establish non-existence of facts.

[10] For a searching critique of modern cosmological theories see Penrose (2004).

A number of points can be made in reply to this objection. First, it may be that, even though value features exist, nevertheless questions of value are inherently more difficult to settle than scientific questions of fact. Second, it may be much more difficult and problematic to set up a team of experts to decide value-questions than it is to set up a team of experts – the scientific community – to decide questions of scientific fact. Third, apart from fundamentalists of various persuasions, our modern world is awash with subjectivism and relativism, doctrines that deny the very possibility of learning about what is of value. In such a cultural climate, it is hardly surprising that people fail to learn about what is of value, and do not know how to resolve conflicting views about what is of value rationally. Finally, the idea that we might one day develop, what we do not have at present, something like a "science" of value is not nearly as absurd as it may at first seem to be. In chapter six, as I have already indicated, I will outline a conception of inquiry rationally designed to help us learn about what is of value, and resolve conflicts in our values and ideals.

Subjectivism and Relativism about what is of value deserve to be rejected, I conclude, along with Dogmatic Objectivism. What is of value does really exist. The God-of-Cosmic-Value does really exist.

One important question remains. What exactly is the connection between the *existence* of value qualities or facts, and the *unique truthfulness* of one system of values? Could one not acknowledge the first, and yet deny the second, and be perfectly consistent thereby? If we interpret *value realism* to be the first doctrine, and *value objectivism* to be the second, the question amounts to this: Does value realism imply value objectivism? Or are the two doctrines logically independent?

Consider the conjecture, indicated above, that living life lovingly is what is of supreme value. And consider a rival doctrine, one which most people will hold to be peculiarly obnoxious: the Nietzschean or fascist doctrine that what is of supreme value is, not love, but strength, power, leadership, the power of the Nation invested in the uniquely significant figure of the Great Leader. If value-features prized by the philosophy of love exist then surely, equally, value features prized by the Nietzschean philosophy must exist as well: Nietzschean qualities of strength, ruthlessness, dominance, brutality. Value realism does not, in itself, render the philosophy of love true and the Nietzschean philosophy false – as long as both philosophies can be formulated so as not to include false factual statements.

But can they? Could it not be argued that both philosophies must include a statement to the effect that what this philosophy holds to be of supreme value, actually and potentially in existence, really *is* of supreme value, but only one of these statements can be factually *true* – true to what is, in reality, of supreme value? The real existence of what is of value does, in other words, render at least one (and perhaps both) of these clashing philosophies false.

If this argument is valid, it establishes only the *falsity* of one of the philosophies of value. It does not establish that we can know for certain which is false. It provides a rationale for attempting to improve our ideas about what is of value, but very definitely does not justify any claim that this or that view about what is of supreme value is correct.

Does it make sense to talk of what is of value dissociated from what individuals value?

In considering what is of value in the world, much depends on whose interests we take into account, and how. Consider, to begin with, what is of value to one individual, ignoring the interests of others unless these interests are of value to this one person. There is, first, what this person would declare to be of value to him were he to be asked. Second, there is what the person really holds to be of value. Third, there is what the person actually values (which may not be the same as what he believes or says he values). And fourth, there is what really is of value to this person, actually and potentially, in the circumstances of this person's life. This is what we should mean by "what is of value to this person".

But now we may take what is of value to others into account. Let us suppose, to begin with, that we consider what is of value to individuals belonging to a well-defined group, G. We might, to begin with, consider what is of value to the individuals belonging to G irrespective of whether these values clash or not. But then, second, we might consider what is of value to the members of G with all clashes resolved in the best possible way, doing the best justice to what is of value to each of the individual members of G [11]. Third, G might represent some group endeavour, with some aim, and with its own interests. An example is science given the aim, let us suppose, of improving knowledge and understanding of aspects of the universe. What is of value to *science* need not at all be the same as what is of value to *scientists*. Improved knowledge and understanding are of value to *science*; these may well be of value to individual scientists as well, but many scientists will value recognition, gaining a Nobel prize – of no value whatsoever to science itself (except perhaps as a means to motivate the servants of science to serve its interests).

Can G be generalized to become humanity, everyone, all sentient beings? Perhaps. But at this point, things become a bit hazy. Do we include unborn children? This is not a merely academic question. In deciding issues about the environment and global warming, we certainly take into account the interests of the unborn, those who will be alive in 50 to 100 years time. But who exactly are these unborn? How, even in principle, could they be specified – except, perhaps, to say, simply, they are who they will be. The idea of what is of value, actually and potentially, to everyone, born and unborn, all conflicts resolved in the best possible way, clearly faces problems. But it would seem to be this problematic idea that value realism requires.

Does it make sense to speak of what is of value in a way which is *dissociated* from all individuals, all sentient beings? Is there such a thing as what is of value *in itself*, as it were, irrespective of whether it is of value to this or that person? My answer is No. I can allow that aspects of the natural world might be beautiful, and thus of value, even though no one exists to experience and appreciate it (perhaps because the human race has died out); but this appeals to the *possibility* of personal experience. What I do not believe is that something of intrinsic value exists dissociated from all sentient life, actual or possible. The God-of-Cosmic-Value is absolutely essentially linked to conscious, or sentient, life.

---

[11] It does not matter if there are many different, equally good, just ways of supposing conflicts to be resolved, in this way.

How can we make discoveries in this domain of value, learn about what is of value?

As I have indicated towards the end of chapter two, and as I shall spell out in more detail in chapter six, the way to learn about what is of value is to put wisdom-inquiry into practice as we live.

Is there a realm of higher, spiritual values, glimpsed by some in rare moments of mystical inspiration, and experienced by seers and mystics?

Being alive is miraculous. Of course, *how* miraculous, and how aware you are of how miraculous it is, may depend on who you are. A firefly probably does not feel it at all. A mouse might get an occasional sniff of it. Jane Goodall describes Chimps experiencing wonder.[12] You and I experience the miraculousness of existence from time to time but possibly not with as much persistent intensity as William Blake did, or John Keats or the youthful William Wordsworth. But it is there, available potentially for everyone, the quite incredible mystery and miraculousness of existence, of being alive, sentient and conscious. Here, we may think, is the crucible of value, the fire in which it becomes incandescent.

But is that right? Do intense mystical experiences reveal to us the inner essence of the value of things? Or can they? Or should all such experiences be dismissed as wholly uninformative, sometimes perhaps of great emotional value to the person who has such an experience, but otherwise signifying nothing?

I am very much in two minds about this question. What is deeply, profoundly of value in existence has to do, it seems to me, with loving relationships, with friendship, with kindness, with caring for others, with artistic and intellectual endeavour, with the experience of beauty, with joy in being in the world with others, yes, but not necessarily with mystical excesses. These latter, I am inclined to believe, may have more to do with prior long-standing repression, with the sudden volcanic explosion of psychic energy long held in check. Such psychic explosions are not necessarily a sign of insight, of perception, maturity and health. They do not reveal a spiritual reality ordinarily concealed to us.

---

[12] J. Goodall (1971).

On the other hand I have myself, on a number of occasions, had extraordinary, apparently revelatory experiences which have had a major impact on my life, and on my thinking.[13] This happened for the first time like this. A student at University College London doing mathematics, I had recently been informed my grant had been stopped because I had not attended enough lectures. After an attempt to move to philosophy – which failed because I did not have O level Latin – I decided to leave the University. I was in disgrace, my life, it seemed, a disaster. I had spent the evening in question with my sister. We did not then ordinarily communicate very well, but on this occasion we did. We talked, easily and naturally about all manner of things. Walking back to Camden Town tube station afterwards, it occurred to me that I was not entirely the wretch I took myself to be. My parents and sisters, I began to feel, thought well of me. I was loved. And then, as I walked along the pavement towards Camden Town, everything took on a terrifying intensity of significance. What was happening now, this precise concatenation of particular circumstances, these shifting vistas of buildings in the evening air, would never, never, never occur again, in the entire history of the universe. A woman walked along the pavement in the opposite direction to me, on the other side of the street, her high heels clicking as she walked. This infinitely, uniquely precious moment came but once in all of eternity. Then, as I walked, it seemed to me that the sky had become a vast lens concentrating all the love, all the energy, in the universe into me. It began to feel as if an invisible stream of lightning was tearing silently through my lungs. And instead of being extraordinarily uplifting, as it had begun, the experience became terrifying. "What is happening to me?" I asked. I thought it likely I was about to experience an epileptic fit. I did what I could to distract my attention from what I was experiencing, and gradually, over an hour or so, it wore off. One point struck me forcefully afterwards. If I had had the slightest temptation to believe in the existence of God, I would have interpreted what I had been through as direct mystical communion with God – or at least with some spiritual reality infused into the constitution of things.

---

[13] For brief accounts see www.nick-maxwell.demon.co.uk, "About Me" and "Life of Value". See also my (2006a).

Experiences such as this cannot invariably, I think, be dismissed as psychotic episodes, hints merely of possible, nascent schizophrenia. Instead, such experiences can, perhaps, be interpreted much more positively as vividly reminding us of just how miraculous it is to be alive and conscious, able to share friendship and love with fellow human beings.

I cannot debunk such experiences as symptoms, merely, of psychic illness because, three years later, I had such an experience, not for three hours, but for six weeks. My life was transformed. All my work stems from what I thought, felt and experienced during that six weeks in the summer of 1961, as I have explained elsewhere.[14] This very book ultimately stems from that time. A part of the revelation I experienced was the shock of being liberated from Cartesian dualism and encountering for the first time, in the raw as it were, the experiential God-of-Cosmic-Value, both within and without. But it also had to do with a radical reassessment of my aims and ideals in life, and a realization that philosophy should be centrally concerned with such matters. Both themes run through all my subsequent work, and are basic to this book.

How can we best go about enhancing our capacity to realize what is of value, actually and potentially, in our life, as we live, whatever exactly may be of value, and however it may be conceived?

This is the really important question, of course. What I have to say by way of a contribution towards answering it has already been touched upon towards the end of chapter two, where I argued – in effect – that we urgently need new institutions of learning rationally devoted to helping the God-of-Cosmic-Value flourish within the God-of-Cosmic-Power. This argument will be taken up again in chapters six and nine.

---

[14] See note 13.

CHAPTER FIVE

WHAT IS THE GOD OF POWER?
HOW DO WE KNOW IT EXISTS?

In chapter one I declared that the God-of-Cosmic-Power is that impersonal *something* that exists everywhere, eternally and unchanging, throughout all phenomena, and determines (perhaps probabilistically) the way phenomena unfold. It is what corresponds physically to the true unified theory of everything that physics seeks to discover.

But how do we know that the God-of-Cosmic-Power, in this sense, exists? How do we know that what determines what goes on everywhere in the universe has this character? My answer is that it is *science* that has established that the God-of Cosmic-Power exists. In so far as physics has established anything theoretical, it has established this profoundly important point that the God-of-Cosmic-Power exists. This is more firmly established than the truth of any physical theory, however well corroborated by evidence. It is a more firmly established item of scientific knowledge than our very best physical theories – Newtonian theory, quantum theory, Einstein's theories of special and general relativity.

Any scientist reading this book will, at this point, be in a state of apoplectic outrage. Physics cannot possibly have established that "the God-of-Cosmic-Power" exists! This is an untestable proposition, neither verifiable nor falsifiable empirically. It is, in this respect, like the proposition "every event has a cause", which cannot be verified or falsified empirically either. Neither proposition can be a part of scientific knowledge. In order to be an item of scientific knowledge, a proposition must be, at the very least, empirically testable. It must, in addition, have been subjected to empirical testing, and must have met with sufficient empirical success. *In science, no proposition about the world is accepted as a part of scientific knowledge independently of evidence.* The claim that the God-of-Cosmic-Power exists has no empirically testable consequences, and thus can have no empirical support. It does not even meet the first requirement for being a scientific statement.[1]

One day theoretical physicists may formulate a unified theory-of-everything, a theory which unifies all existing fundamental theories in physics. This theory might meet with sufficient success when subjected to empirical tests to be judged to be a part of scientific knowledge. When, or if, this happens, there might be grounds for declaring that science has established that the God-of-Cosmic-Power exists. But this has not yet happened. It may never happen.

Many theoretical physicists believe that string theory will one day turn out to be the correct theory-of-everything. String theory holds that fundamental particles – electrons, quarks and the rest – are really tiny strings vibrating in ten or eleven dimensions of space-time. We only notice four dimensions of space-time because the other six or seven or curled up into a minute ball, too small to be observed. String theorists admit, however, first, that string theory is not yet a properly formulated physical theory. And they admit that no successful empirical predictions have been forthcoming. String theory is in the same category as the statement that the God-of-Cosmic-Power exits. It is a *metaphysical* theory – that is, a theory neither verifiable nor falsifiable empirically. One day it may meet with empirical success and may become a part of scientific knowledge. But today it is no more than speculative metaphysics.

A New Conception of Science

This counter-argument of the outraged scientist is entirely valid – just as long as the conception of science on which it is based is valid. But it is not. The claim that physics accepts theories on the basis of evidence, *no proposition about the world being accepted as a part of scientific knowledge independently of evidence*, is just plain false. Physics only ever accepts theories that are *unified* or *explanatory*, even though endlessly many empirically more successful disunified, non-explanatory rival theories are always

---

[1] Karl Popper is famous for arguing that a statement or theory, in order to be scientific, must be empirically falsifiable: see K. Popper, 1959, *The Logic of Scientific Discovery*, Hutchinson, London. Most scientists endorse this view (although some would hold that theories can be verified too, something Popper denies).

available. This means – as we shall see in greater detail below – that physics implicitly accepts permanently the very substantial thesis that the universe is such that no seriously disunified, non-explanatory theory is true. Physics accepts, in other words, that the universe is more or less unified, i.e. that it is such that explanations for phenomena exist to be found. This substantial, metaphysical thesis is accepted permanently as a part of scientific knowledge *independently of evidence*, even, in a certain sense, *in violation of evidence.*

What this means is that we need to adopt a whole new conception of science, very much at odds with current orthodoxy. Natural science, far from being dissociated from religion and metaphysics is, on the contrary, at the most fundamental level, a religious and metaphysical quest in its own right. At the most fundamental level, science accepts, as a basic tenet of scientific knowledge, a metaphysical – that is, an empirically untestable – thesis about the nature of the universe. This metaphysical thesis specifies what it is in nature that is responsible for events occurring as they do – what it is, in other words, that is *in charge* of the universe. A central task of natural science is to *improve* this metaphysical thesis – get it to reflect, more and more accurately, the nature of that which is ultimately responsible for determining how events unfold. Science does this by accepting that thesis which seems to be the most empirically fruitful, in the sense of supporting the most empirically successful scientific research programme. Viewed from this standpoint, we engage in empirical scientific research in order to improve a basic, empirically untestable, metaphysical thesis about the nature of the universe – one which specifies what it is in existence that is ultimately responsible for events occurring as they do. It is in this sense that theoretical physics is an empirical quest for the nature of God – the nature of the God-of-Cosmic-Power. Nevertheless, despite its empirical character, the whole enterprise of natural science rests on an article of faith: that something does exist which determines how events unfold, and which explains why events occur as they do – the universe being ultimately comprehensible, in some way or other.

Underlying Theoretical Unity in Nature

Despite the above, it may seem quite impossible that we should be able, today, to capture the ultimate nature of the physical universe in a metaphysical thesis that stands a reasonable chance of being *true*. This is in part because a *metaphysical* thesis, being empirically untestable, seems to provide no means for empirical assessment. (This will be rebutted below.) But a far more serious reason for doubt arises from the history of physics. Ever since modern science began, metaphysical ideas about how the universe is physically comprehensible, associated with physics, have changed dramatically, again and again. In the 17th century, there was the idea that everything is made up of minute, rigid corpuscles that interact only by contact. This gave way, in the 18th and 19th centuries, to the idea that everything is made up of point-particles that have mass but no size. These point-particles were thought to interact by means of forces at a distance, alternatively attractive and repulsive, and becoming like gravitation at sufficiently large distances. This in turn gave way, after Faraday and Einstein in the late 19th and early 20th centuries, to the idea that everything is made up of a unified, classical field. A physical field is a physical entity, like the magnetic or electric field, which is spread out smoothly in space. Especially dense regions of the field appear to be particles. The field interacts with itself, so that dense regions – which appear to be particles – are pushed about by the surrounding field.[2] This in turn gave way to the idea that everything is made up of mysterious quantum objects, exhibiting both wave and particle features, which in turn transmuted into the idea of current string theory: the universe is made up of minute quantum strings oscillating in ten or eleven dimensions of space-time.

---

[2] An influential view, half-way between the point-particle and field views is the idea that point-particles are embedded in a field, which is created by, and which acts on, the point-particles. This point-particle/field view, however, faces the severe problem of an infinitely strong interaction between point-particle and field, because of the point-like character of the particles, and because the particles both create, and are acted upon, by the field. The original point-particle view does not encounter this problem because each point-particle creates its own field of force *which acts only on other particles*.

125

Given this dramatically changing sequence of ideas, what hope is there that any idea formulated today will turn out to be the final truth, a permanent item of scientific knowledge for the rest of time?

Metaphysical ideas about the ultimate nature of the physical universe have changed so dramatically because of dramatic revolutions in theoretical physics. Newton's theory of gravitation which, it seems, postulates the existence of a force acting at a distance (for example between the earth and the sun) makes nonsense of the corpuscular hypothesis (which holds all interactions are by contact). The idea of point-particles interacting by means of forces at a distance arose instead, in response to Newton's contribution. Subsequently, Faraday and James Clerk Maxwell developed the classical theory of the electromagnetic field. This clashes with Newtonian theory and the idea of the point-particle. It requires the existence of a new kind physical entity, the classical field. This in turn became implausible after the advent of quantum theory. This theory led eventually to the current idea: everything is made up of minute quantum strings.

It is, in short, these dramatic revolutions in theoretical physics which have forced physicists to revise their metaphysical ideas about the nature of the physical universe. That science advances by means of intermittent revolutions in theory is an idea that has been made popular by Thomas Kuhn as a result of his book *The Structure of Scientific Revolutions* (1962). This Kuhnian view of how science advances would seem to make it even more implausible to suppose that we could, today, specify a metaphysical thesis about the ultimate nature of the physical universe which will stand as true for all time. How could this be done, when we take into account the dramatic revolutions that have taken place, in metaphysical ideas, and in fundamental physical theory, since modern science began? Why should future physics be any different from past physics?

But these arguments, apparently so persuasive, all fall to the following consideration. All revolutionary new theories in physics have one decisive feature in common. The new theory succeeds, dramatically, in *unifying* what beforehand appeared to be disparate phenomena. Before Galileo and, above all, Newton, terrestrial phenomena were believed to be quite different from astronomical phenomena, in line with the then prevailing view of Aristotle.

Newtonian theory (his laws of motion and law of gravitation) apply equally to terrestrial and astronomical phenomena (the motion of a thrown stone *and* the motion of the planets round the sun). Before Faraday and Maxwell, electricity, magnetism and light seemed to be three quite distinct phenomena. Maxwell's theory of the electromagnetic field unifies electricity and magnetism, and reveals that light is waves in the electromagnetic field. Before quantum theory, endlessly many diverse laws were required to specify the myriad diverse properties of elements and chemical compounds. After the advent of quantum theory, and the theory of atomic structure, these myriad diverse laws, corresponding to myriad diverse properties, become, in principle unified by means of the very few basic physical properties of sub-atomic particles, electrons, neutrons and protons, and quantum theory. Likewise, before the advent of Einstein's theories of relativity, space, time and gravitation seemed three quite distinct entities. They are unified by Einstein's theory of general relativity, according to which gravitation is a feature of curved space-time. And yet again, the so-called standard model, the current quantum theory of fundamental particles and fields, brings far greater unity to fundamental particles and forces than appeared to exist before the advent of the theory.

In short, just that which seems to rebut the idea that there is a permanent metaphysical thesis inherent in the history of physics – namely a succession of revolutionary new theories which contradict their predecessors – actually provides dramatic support for the idea. Every revolutionary physical theory brings greater unity to theoretical physics. It is precisely the revolutionary theoretical developments which reveal the persistence of the idea: there is underlying theoretical unity in nature. The history of physics, despite its theoretical ruptures, or rather because of them, actually endorses the point I made above (and will further develop below): physicists only ever accept *unifying* theories, even though disunified rivals can always be concocted which would fit the evidence even better; this means physics persistently accepts a metaphysical thesis about the nature of the universe, throughout theoretical revolutions (of past and future), namely that there is underlying unity in nature. In order to recognize what persists through revolutions, we need to formulate an idea more abstract than that of the corpuscle, point-particle, field or string – the idea, namely, of some kind of underling theoretical *unity* in nature.

At once a number of questions arise. (1) What reason can there possibly be for holding that this metaphysical thesis is true, or at least sufficiently likely to be true to be accepted as a fundamental component of scientific knowledge? (2) Why is it necessary for science to make this assumption? (3) What precisely does this metaphysical thesis, that is supposed to be presupposed by science, assert? (4) How can science set about *improving* this metaphysical thesis concerning the comprehensibility of the universe? What procedures or methods can be adopted by science to give the best chance of adopting the best available version of this presupposed metaphysical thesis? (5) How can *empirical* considerations affect what metaphysical thesis is accepted given that these theses are *metaphysical*, and thus empirically untestable? (6) Does not this view of science, which holds that science rests on *an article of faith*, an *untestable, metaphysical assumption*, amount to a dreadful betrayal of scientific rigour and objectivity? (7) What does it *mean* to assert that one metaphysical thesis is an *improvement* over another, in that it reflects, more accurately, the nature of that which is ultimately responsible for determining how events unfold? What does it mean to say of one thesis that it is closer to the truth than another? (8) How can it be possible for what exists at one instant to determine what exists next, given David Hume's decisive arguments which show that this is not possible? (9) Does "The God-of-Cosmic-Power exists" really amount to a meaningful, substantial assertion? (10) Does not quantum theory refute the thesis that only unified theories are accepted in physics?

In what follows, I shall answer these questions in turn. If what follows starts to get too technical, read the next section at least, look at diagram 4, which summarizes the basic point, and go on to chapter six.

Standard versus Aim-Oriented Empiricism
We have before us two rival views of science which, in past discussions of these issues I have called *standard empiricism* and

aim-oriented empiricism.[3]

Standard empiricism is the view taken for granted by the outraged scientist, mentioned above. It holds that i*n science, no proposition about the world is accepted as a part of scientific knowledge independently of evidence.* In deciding whether or not to accept a theory, scientists may legitimately be influenced by the simplicity, the unity, or the explanatory character of the theory, but this influence must not operate in such a way that scientists in effect, surreptitiously perhaps, assume that the universe itself is simple, unified, or comprehensible. This view of standard empiricism is the current orthodoxy. It is widely taken for granted, by scientists and non-scientists alike. It is an explicit or implicit component of many views about the nature of science, from the views of the logical positivists to those of Karl Popper, Thomas Kuhn, and others.[4] If valid, it makes a nonsense of the claim that science has established that the God-of-Cosmic-Power exists. But standard empiricism is not valid. It is, as we shall see in more detail, untenable.

Aim-oriented empiricism is based on the argument that physicists, in only ever accepting unified, explanatory theories even though endlessly many empirically more successful *disunified* rivals are always available, thereby surreptitiously make a big assumption about the universe. The universe is such that no seriously disunified theory is true. It is more or less physically comprehensible. Some kind of underlying unity in the physical universe – which is what the God-of-Cosmic-Power is – does actually exist. Natural science is more rigorous if this substantial, problematic and implicit assumption is made explicit, so that it can be improved as an integral part of the scientific enterprise. Given this view of aim-oriented empiricism, it makes perfect sense to say science has established that the God-of-Cosmic-Power exists. To this extent, science is a religious endeavour. It requires, as an article of (rational) faith, that the God-of-Cosmic-Power exists, and seeks to discover its precise nature.

---

[3] See my (1974); (1976a) or 2nd edition (2009); (1984) or (2007a); (1998); (2004a), especially the appendix.
[4] For the point that Popper, Kuhn and Lakatos all held versions of standard empiricism see my (2005a). For grounds for holding scientists accept standard empiricism see my (2004a), ch. 1.

I now answer in turn the above nine questions. In doing so, I spell out what aim-oriented empiricism amounts to in more detail, and give the argument for rejecting standard empiricism and adopting aim-oriented empiricism in its stead as the new scientific orthodoxy.

(1) What reasons are there for holding that the metaphysical thesis, supposedly accepted by science, is true?

The answer is that, what reasons there are, are extremely flimsy. All our knowledge, including all our scientific knowledge, as Karl Popper for one has tirelessly argued, is conjectural. [5]   No scientific theory can be proved or verified by evidence. All we can hope to do in science is *refute* a theory empirically. But it is this which makes scientific progress possible.   As a result of falsifying a theory empirically, we are forced to try to think up a better theory, one that successfully predicts everything the refuted theory predicts, successfully predicts the phenomena that falsified the earlier theory, and successfully predicts new phenomena.   Subjecting our theoretical scientific guesses to severe empirical testing uncovers falsity in what we have accepted as scientific knowledge, and forces us to revise and improve our conjectural scientific knowledge. It is by means of this process of developing imaginative testable guesses which are then subjected to severe attempted falsification that science makes progress.[6]

And the point can be generalized. In tackling a problem in any field of human endeavour, in order to give ourselves the best chance of success we need to make imaginative guesses as to what the solution to the problem might be; we need then to subject these

---

[5] See Popper (1963); and, at a more technical level, (1959).

[6] I am about to argue, I hope it is clear, that this Popperian picture of how and why science makes progress needs to be revised. Physics only ever puts forward falsifiable conjectures that are *explanatory* (even though empirically more successful non-explanatory conjectures are always available).   This means physics makes the permanent, implicit, metaphysical assumption that the universe is such that explanations exist to be found – the universe is more or less physically comprehensible, in other words. This implicit assumption needs to be made explicit within science so that it can be critically assessed in an attempt to improve it.

guesses to severe criticism (especially when we attempt to put them into practice), in the hope of discovering the best available solution.

The big metaphysical assumption of science – that the universe is more or less physically comprehensible – is, like our very best accepted scientific theories, a conjecture. As in the case of our best scientific theories, there are no arguments which demonstrate that the assumption is true, or likely to be true.

Despite this, there are quite strong arguments for holding that this conjecture needs to be accepted as a rather secure item of theoretical scientific knowledge, an item that is more secure, indeed, than our best scientific theories. This takes us straight to our next question.

(2) Why is it necessary for science to make this assumption?

I have already indicated why science must assume that the universe is more or less physically comprehensible, whether this is acknowledged or not. I now spell out the argument in a little more detail.[7]

In physics, two considerations govern acceptance of a theory: empirical success[8], and the *unified* or *explanatory* character[9] of the theory in question. A disunified, "patchwork quilt" theory, made up of a number of different laws which apply to different ranges of phenomena, would never be accepted in physics however empirically successful it might prove to be if considered.[10] In

---

[7] In what follows I summarize arguments spelled out in greater detail in the works referred to in notes 3 and 4.

[8] What does "empirical success" mean here, given the earlier admission that no theory can be empirically verified? It means that the theory in question has been subjected to severe attempted empirical falsification, and (ideally) has survived, unrefuted. It means, in addition, that the theory successfully predicts more phenomena than any (unified) rival theory.

[9] We may take "unified" and "explanatory", in the present context, to be equivalent. What "unity" of theory means here will be clarified further in the text below.

[10] This is not quite correct. Quantum theory is an accepted theory; but it is just such a "patchwork quilt" theory, as we shall see in section (10) below. We shall also see in section (10) that this case of quantum theory does not refute the general argument.

short, in physics, unified (or explanatory) theories are only ever accepted, even though it is always the case that endlessly many disunified, "patchwork quilt" rival theories can be concocted which would be even more empirically successful if considered. (A physical theory is unified if it attributes *the same* laws to the range to phenomena to which it applies.)

Now comes the crucial point. In persistently accepting *unified* theories only, even though endlessly many empirically more successful *disunified* rival theories are always available, physics in effect makes a big persistent assumption: all seriously disunified theories are false. Physical laws governing phenomena are unified. The universe is such that the yet-to-be-discovered true theory of everything is more or less unified, or explanatory

This means standard empiricism is false. Persistent acceptance of unified theories only, even though empirically more successful disunified rivals are available, means that physics does accept one big assumption about the nature of the universe permanently, entirely independently of evidence – even, in a sense, at odds with the evidence. Standard empiricism is untenable, and must be rejected.

If scientists only accepted theories that postulate atoms, and persistently rejected theories that postulate different basic physical entities such as fields – even though many field theories can easily be, and have been, formulated which are even more empirically successful than the atomic theories – the implication would surely be quite clear. Scientists would in effect be assuming that the world is made up of atoms, all other possibilities being ruled out. The atomic assumption would be built into the way the scientific community accepts and rejects theories – built into the implicit *methods* of the community, methods which include: reject all theories that postulate entities other than atoms, whatever their empirical success might be. The scientific community would accept the assumption: the universe is such that no non-atomic theory is true.

Just the same holds for a scientific community which rejects all disunified rivals to accepted theories, even though these rivals would be even more empirically successful if they were considered. Such a community in effect makes the assumption: the universe is such that no disunified theory is true (unless implied by a true unified theory).

The alert reader may have noted a weak link in this argument. What grounds do we have for holding that endlessly many empirically more successful but disunified rivals to any accepted physical theory can always be concocted? Well, consider any accepted physical theory, T say. T might be Newtonian theory, quantum theory, or Einstein's theory of general relativity. However empirically successful and well established T may be, there will always be a range of phenomena, B, that the theory

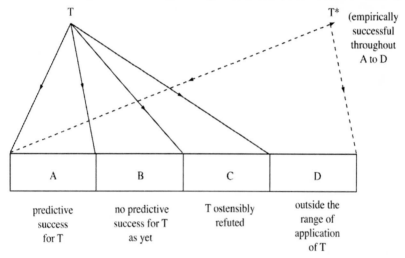

Diagram 3
Empirically Successful, Disunified Rival Theories

cannot (as yet) predict, a range of phenomena, C, that ostensibly refute T, and a range of phenomena, D, which lie outside the predictive range of T altogether: see diagram 3. This means a rival theory, T* can be concocted which meets with empirical success throughout the range of phenomena A to D. T* is just like T for phenomena A, while for phenomena B, C and D, T just incorporates the empirical laws that describe these phenomena. Unquestionably, T* is empirically more successful than T. T* successfully predicts everything T predicts; it predicts phenomena that T fails to predict (B and D), and is not refuted where T is

133

ostensibly refuted (C). Endlessly many horribly disunified but empirically more successful rivals to T, like T*, may be concocted along these lines by, for example, modifying T for phenomena for which it has not yet been tested, and then carrying through a procedure similar to the one indicated, used to concoct T*.

The conclusion is inescapable. In persistently accepting unified theories, like T, in preference to endlessly many empirically more acceptable but disunified rivals, like T*, physics makes a persistent assumption about the nature of the universe, independent of empirical considerations, to the effect that it is such that no seriously disunified theory is true. Standard empiricism, presupposed by the above argument against the thesis that physics has established that the God-of-Cosmic-Power exists, really is untenable and must be rejected.

That physics makes metaphysical or cosmological assumptions really ought not to be thought surprising or controversial. Even our most mundane items of common sense knowledge make such assumptions. In so far as I "know" I can, for the next two minutes, continue to sit in the chair I am sitting in as I type these words, I "know" something about the entire cosmos. I know that nowhere in the cosmos is there a vast conflagration occurring which will spread almost instantaneously to engulf and destroy the earth, my chair, and me, in the next two minutes.[11] If even meagre, almost trivial items of common sense knowledge such as this contain (some) knowledge about the entire cosmos, it should not surprise us that the theories of physics, embodying vastly greater amounts of knowledge, should turn out to contain some knowledge about the entire cosmos too.

---

[11] If I know p, and p implies q (and I know it) then I know q too. Me continuing to sit in my chair (p) implies no conflagration is occurring anywhere which will destroy me (q) – and I know p implies q. Hence knowing I can continue to sit in my chair (p) means I know no conflagration is occurring anywhere which will destroy me (q). Even our most trivial common sense knowledge has a cosmological dimension to it.

(3) What precisely does the metaphysical thesis that is presupposed by science, assert?

I have already indicated what the answer to this question is. In persistently accepting unified theories only, physics implicitly assumes that the universe is such that no seriously disunified theory is true. But what exactly does this mean? How seriously disunified must a theory be to be unacceptable? And if a disunified theory is, in effect, two or more distinct theories, postulating distinct laws, stuck artificially together, how do we distinguish the case of *one* disunified theory made up of *three* distinct "patchwork quilt" parts stuck arbitrarily together, from the case of *three* unified theories? Cannot disunified theories be turned into unified theories by the simple dodge of increasing the number of theories in the way indicated?

The way to avoid this dodge is to consider the degree of disunity of *all* fundamental theories in physics. We consider T (let us call it), made up of all current fundamental physical theories which, together, suffice to apply to all known possible physical phenomena. (If there are phenomena which do not, as yet, have a theory that applies to them, then we include in T the experimental laws governing these phenomena.) T today consists of Einstein's theory of general relativity, and the quantum theory of fundamental particles and the forces between them – the so-called "standard model". We can at once say that T today is disunified at least to degree 2, because it is made up two very different theories stuck artificially together, Einstein's theory of general relativity which applies to gravitation and to space and time in the large, and the standard model, a quantum theory which applies to fundamental particles (such as electrons and quarks) and the forces between them (such as the electromagnetic force). Actually, T today is much more seriously disunified than this because the standard model is itself seriously disunified, in that it is about three different forces, and many different particles with different properties such as their values of charge and mass.[12]

---

[12] T is even more disunified if dark matter, for which there is, at present no theory, really does exist. Dark matter has revealed its presence so far only through its gravitational effects. Galaxies rotate so fast that, unless there is invisible dark matter in addition to the observed matter, the galaxies would fly apart. There is always the faint possibility, however,

How, precisely and in general, do we judge how disunified a physical theory, T, is? The crucial step is to attend, not to the theory itself (its form, it axiomatic structure, or anything of that kind), but to *what the theory asserts about the world.* And for full unity we require that T asserts that *the same laws* govern all the phenomena to which T applies. (For full theoretical unity, it does not matter if the linguistic form of the laws changes as we move from one range of phenomena to another, as long as *what is asserted* remains the same.) If the range of phenomena to which T applies can be split up into N sub-domains, different laws governing the phenomena in each of these sub-domains, then T is disunified to degree N (where N is some integer). If N = 1, then the theory in question, T, is fully unified.

If the true "theory of everything", T, is unified, with N = 1, then the physical universe is made up of two parts or aspects: **V** and **U**. **V** stands for *that which varies.* It is made up of everything in the universe which varies or changes, from time to time and place to place. In a universe made up of particles, **V** would be the variable positions and velocities of the particles. Or, if the universe is made up of a physical field, **V** is the varying intensity of the field, from moment to moment, and place to place. **U**, on the other hand, stands for *that which is unchanging* – that which is the same everywhere, at all times and places, throughout all possible phenomena, and determines the way **V** does vary. If the universe is made up of a physical field, then **U** would be that property or aspect of the field that is the same, everywhere, throughout all change, and determines the way the variable aspects of the field do vary, with the passage of time. It is **U** that is the physical God-of Cosmic-Power.

The nature of **U** is specified by the true, unified, physical theory of everything, T. **V** is specified by so-called initial conditions: these specify, at any given instant, the instantaneous, variable physical state of an isolated bit of the universe.

---

that Newton's and Einstein's theories are seriously wrong at great distances, and the correct theory of gravitation would predict that the galaxies hold themselves together without additional unseen dark matter – dark matter being non-existent.

U determines the way V varies as follows. Let there be a physical system isolated from all external influences (which could be the entire universe). Then, given a precise specification of V at some definite instant, and given a precise specification of U, by the physical theory T, subsequent instantaneous, variable states of the system, V, are logically entailed (as long as the system remains isolated). Given that the initial variable state at time t_o is Vt_o, and the state at a subsequent time is Vt, then we might write the way in which U determines the way in which V varies like this:
$U + Vt_o \rightarrow Vt$.

This presupposes determinism. It may well be, however, that nature is fundamentally probabilistic, rather than deterministic. In this case $U + Vt_o$ would determine many, even perhaps infinitely many, different possible Vt, and would assign a probability of occurrence to each. In a fundamentally probabilistic universe, the present only determines the future probabilistically.

If the universe is disunified to degree N, then there is no one invariant U running through all phenomena, actual and possible. Instead, there are N distinct Us – we may call them $U_1, U_2 \ldots U_N$ – each one determining the way V varies for some distinct range of phenomena. In this case the God-of-Cosmic-Power is $U_1 + U_2 + \ldots + U_N$.

But now there is a complication – or rather, a refinement. Given a theory that is disunified to degree N > 1, the question can arise as to *how different, in what way different*, are laws in one range of phenomena from laws in another range of phenomena – or how different $U_1, U_2 \ldots U_N$ are from one another. Some ways in which sets of laws can differ, one from the other, can be much more dramatic, much more serious, than other ways. This gives rise to different *kinds* of disunity, some being much more serious than others.

Here are five different ways in which dynamical laws can differ for different ranges of phenomena, and thus five different *kinds* of disunity.

137

Imagine that the physically possible phenomena, to which the theory T applies, are spread out before us. A phenomenon, here, is a physical system, a group of interacting particles for example, located somewhere in space and evolving in time. We are concerned with physical phenomena that are *possible*, not just with actual phenomena. For simplicity, let us assume that N = 2 throughout. The phenomena to which T applies split up into two regions, one set of laws applying in one region, a *more or less* different set of laws applying in the other region. (N might be any integer, 6 say, or 42, the bigger it is, so the greater the degree of disunity of T, the theory in question. We take N = 2 to keep things as simple as possible for expositional reasons.)

There are now the following possibilities to consider.

(1) The laws specified by T are different in different regions of space, or different at different times. This is the most serious kind of disunity. An example would be a "patchwork quilt" version of Newton's law of gravitation, which asserts that gravitation is an attractive force up to midnight tonight, and a repulsive force after midnight. (Here, T is disunified to degree N = 2 in a type (1) way.)

(2) T differs in N distinct ranges of physical variables other than position or time. Example: $F = Gm_1m_2/d^2$ for all bodies except for those made of gold of mass greater than 1,000 tons in outer space within a region of 1 mile of each other, in which case $F = Gm_1m_2/d^4$. Here, T is disunified to degree N = 2 in a type (2) way.

(3) T is unified except that it postulates N distinct, spatially localized objects, each with its own unique dynamic properties. Example: T asserts that everything occurs as Newtonian theory asserts, except that there is one object in the universe, of mass 8 tons, such that, for any two bodies both up to 8 miles from the centre of mass of this object, the force of gravitation between the two bodies is repulsive rather than attractive. The object only interacts by means of gravitation. Here, T is disunified to degree N + 1 = 2, in a type (3) way.

(4) T postulates N distinct forces. Example: T postulates particles that interact by means of Newtonian gravitation; some of these also interact by means of an electrostatic force $F = Kq_1q_2/d^2$, this force being attractive if q_1 and q_2 are oppositely charged, otherwise being repulsive, the force being much stronger than gravitation. Here, T is disunified to degree N = 2 in a type (4) way.

(5) T postulates one force but N distinct kinds of particle. Example: T postulates particles that interact by means of Newtonian gravitation, there being three kinds of particles, of mass m, 2m and 3m. Here, T is disunified to degree N = 3 in a type (5) way.

(1) to (5) are to be understood as accumulative, so that each presupposes N = 1 as far as its predecessors are concerned.[13]

These five facets of disunity all exemplify, it should be noted, the same basic idea: disunity arises when *different* dynamical laws govern the evolution of physical states in different ranges of possible phenomena to which the theory T applies. Thus, if T postulates more than one force, or kind of particle, then in different ranges of possible phenomena, different force laws will operate. In one range of possible phenomena, one kind of force operates, in another range, other forces operate. Or in one range of phenomena, there is only one kind of particle, while in another range there is only another kind of particle. The five distinct facets of unity, (1) to (5) arise, as I have said, because of the five *different* ways in which content can vary from one range of possible phenomena to another, some differences being *more* different than others.

Let me emphasize once again that the above five facets of unity all concern the *content* of a theory, and not its *form*, which may vary drastically from one formulation to another. One might, for example, split space up into N regions, and introduce special terminology for each region so that Newton's laws look very different as one goes from one spatial region to another. Thus, for one spatial region one might choose to write d^2 as "d^6", even though "d^6" is interpreted to assert d^2. As one goes from region to region, the *form* of the theory, what is written down on paper, varies dramatically. It might seem that this is a theory disunified to degree N in a type (1) way – the most serious kind of disunity of all. But as long as *what is asserted, the content,* is the same in all spatial regions, the theory is actually unified in a type (1) way, with N = 1.

It deserves to be noted in passing that this solution to the problem of what it means to say of a theory that it is unified also solves the problem of what it means to say of a theory that it is *explanatory*. In

---

[13] Elsewhere I have spelled out further aspects or kinds of disunity that involve *symmetry*: see my (1998), ch. 4; (2004a), appendix, section 2; (2007a), ch. 14, section 2.

order to be explanatory, a theory must (a) be unified and (b) of high empirical content.[14]

The 5 varieties of theoretical unity, encapsulated in (1) to (5) above, can be collapsed into one basic requirement:

Requirement for Unity: T is disunified to degree N if, given all phenomena predicted by T to be possible (i.e. such that the phenomena occur in accordance with T), then there are N regions in which what T asserts in any one region is different from all the other regions. For theoretical unity, we require that N = 1.

One proviso needs to be added. As we range through the possible phenomena predicted by T, we always consider only the least part of T that is required to predict the phenomena in question. If T postulates two kinds of particle, A and B, then in all those possible phenomena consisting only of particles A, only that part of T is specified which predicts how particles A interact and evolve; the part of T that predict how particles B interact, and how particles A and B interact, is excluded (if it can be excluded). We consider, always, the least possible content of T required to predict the evolution of the phenomenon in question. If the true theory of everything, T, is disunified to degree N, so that that which determines how V varies is made up of N distinct parts, U_1, U_2, . . . U_N, then we require that each distinct part of T, namely T_1, T_2, . . . T_N specifies precisely the physical nature of each corresponding U_1, U_2, . . . U_N, no more, and no less.

It deserves to be noted that if T, otherwise unified, fails to unify matter and space-time, so that one possible physical state is empty space-time, with all matter removed, then T is not fully unified. That part of T specifying empty space-time *and no more*, will be different from that part of T specifying how matter evolves in space-time. The above *Requirement for Unity* demands, for full unity, that matter and space-time are unified – so that matter is an aspect of space-time, or space-time is an aspect of matter, or both are aspects of some third unified entity.[15]

---

[14] I have sketched an account of what it is for a theory to be *unified*, but have not said anything about *simplicity*. For that, see my (1998), ch. 4, section 16; (2004a), pp. 172-4. It is a great success of the theory that it sharply distinguishes the two notions of unity and simplicity.

[15] For further details concerning the account of unity of theory given

(4) What should the metaphysical presupposition of science be? How can it be *improved*?

It has been established that science accepts, as a part of scientific knowledge, that the universe is such that no seriously disunified physical theory is true. And what it means to say of a physical theory that it is "disunified" has been clarified. But this still leaves open the question of precisely what metaphysical thesis science ought to accept (granted that *some* metaphysical thesis is inevitably implicitly accepted as a result of the persistent acceptance of unified theories only, even though empirically more successful disunified rivals can always be concocted).

There is a range of options. One possibility would be to say that physics ought to accept, rather modestly, at any given stage, that metaphysical thesis which asserts that the universe is such that the true theory of all physical phenomena is at least somewhat more unified than the then accepted body of fundamental physical theory (at present general relativity and the standard model). A much more immodest possibility is that physics should accept that metaphysical thesis which asserts that the true theory of all physical phenomena is wholly unified, with $N = 1$. There are even more immodest possibilities, as we have seen, such as string theory.

Science must make some kind of choice. The choice that is made will have a profound affect on scientific progress, on our success or failure in discovering what kind of universe this is. This is because the choice that is made affects what physical theories are sought, and accepted. Making a bad choice will sabotage scientific progress. Science only got going in the 17th century because Galileo (and others) made a rather good choice of basic metaphysics, which led to the adoption of fruitful methods. But choosing is fraught with uncertainty, and almost inevitable error. We are almost bound to make the wrong guess about the ultimate nature of the universe. Metaphysical theses, being untestable, cannot be subjected to straightforward empirical assessment. How, then, is science to choose? And how can it best set about improving its choice?

In order to solve these problems – as I have argued at length elsewhere – we need to reject the current orthodox conception of

---

here, see works referred to in notes 13 and 14.

science of *standard empiricism* and adopt instead the more rigorous conception of science, already mentioned, that I have called *aim-oriented empiricism*[16] The basic idea of aim-oriented empiricism is that we need to see physics (and natural science more generally) as making not one, but a hierarchy of assumptions concerning the unity, comprehensibility and knowability of the universe, these assumptions becoming less and less substantial as one goes up the hierarchy and thus becoming more and more likely to be true, and also becoming such that their truth is increasingly a requirement for science, or the acquisition of knowledge, to be possible at all: see diagram 4. The idea is that in this way we separate out what is most likely to be true, and not in need of revision, at and near the top of the hierarchy, from what is most likely to be false, and most in need of criticism and revision, near the bottom of the hierarchy. Evidence, at level 1, and assumptions high up in the hierarchy, are rather firmly accepted, as being most likely to be true (although still open to revision): this is then used to criticize, and to try to improve, theses at levels 2 and 3 (and perhaps 4), where falsity is most likely to be located.

At the top there is the relatively insubstantial assumption that the universe is such that we can acquire some knowledge of our local circumstances. If this assumption is false, we will not be able to acquire knowledge whatever we assume. We are justified in accepting this assumption permanently as a part of our knowledge, even though we have no grounds for holding it to be true. As we descend the hierarchy, the assumptions become increasingly substantial and thus increasingly likely to be false. At level 5 there is the rather substantial assumption that the universe is comprehensible in some way or other, the universe being such that there is just one kind of explanation for all phenomena. At level 4 there is the more specific, and thus more substantial assumption that the universe is *physically* comprehensible, it being such that there is some yet-to-be-discovered, true, unified "theory of

---

[16] For a detailed exposition and defence of aim-oriented empiricism see my (1998); (2004a), ch. 1 and Appendix; or (2007a), ch. 14). For earlier accounts of aim-oriented empiricism see my (1974); (1976a); (1984); and (1993a). See also my (2002b); (2005a); (2010a); and (2010b).

everything" (with $N = 1$). This thesis may be called *physicalism*: it asserts that the God-of-Cosmic-Power exists. At level 3 there is the even more specific, and thus even more substantial assumption that the universe is physically comprehensible in a more or less specific way, suggested by current accepted fundamental physical theories. Examples of assumptions made at this level, taken from the history of physics, include those already mentioned, from the 17th century corpuscular hypothesis to today's string theory. Given

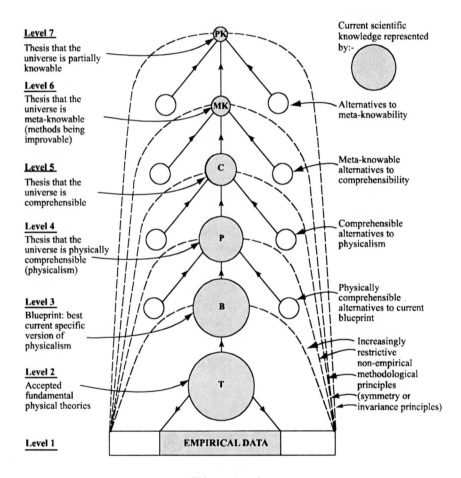

Diagram 4
Aim-Oriented Empiricism

this historical record of dramatically changing ideas at this level, and given the relatively highly specific and substantial character of successive assumptions made at this level, we can be reasonably confident that the best assumption available at any stage in the development of physics at this level will be false, and will need future revision. At level 2 there are the accepted fundamental theories of physics, currently Einstein's theory of general relativity and the so-called "standard model", the quantum theory of fundamental particles and the forces between them. Here, if anything, we can be even more confident that current theories are false, despite their immense empirical success. This confidence comes partly from the vast empirical content of these theories, and partly from the historical record. The greater the content of a proposition the more likely it is to be false; the fundamental theories of physics, general relativity and the standard model, have such vast empirical content that this in itself almost guarantees falsity. And the historical record backs this up; Kepler's laws of planetary motion, and Galileo's laws of terrestrial motion are corrected by Newtonian theory, which is in turn corrected by special and general relativity; classical physics is corrected by quantum theory, in turn corrected by relativistic quantum theory, quantum field theory and the standard model. Each new theory in physics reveals that predecessors are false. Indeed, if the level 4 assumption of aim-oriented empiricism is correct, then all current physical theories are false, since this assumption asserts that the true physical theory of everything is unified, and the totality of current fundamental physical theory, general relativity plus the standard model, is notoriously disunified.

Finally, at level 1 there are accepted empirical data, low level, corroborated, empirical laws.

In order to be acceptable, an assumption at any level from 6 to 3 must (as far as possible) be compatible with, and a special case of, the assumption above in the hierarchy; at the same time it must be (or promise to be) empirically fruitful in the sense that successive accepted physical theories increasingly successfully accord with (or exemplify) the assumption. At level 2, those physical theories are accepted which are sufficiently (a) empirically successful and (b) in accord with the best available assumption at level 3 (or level

144

4). Corresponding to each assumption, at any level from 7 to 3, there is a methodological principle, represented by sloping dotted lines in diagram 4, requiring that theses lower down in the hierarchy are compatible with the given assumption.

When theoretical physics has completed its central task, and the true theory of everything, T, has been discovered, then T will in principle (not in practice) successfully predict all empirical phenomena at level 1, and will entail the assumption at level 3, which will in turn entail the assumption at level 4, and so on up the hierarchy. As it is, physics has not completed its task, T has not (yet) been discovered, and we are ignorant of the nature of the universe. This ignorance is reflected in clashes between theses at different levels of aim-oriented empiricism. There are clashes between levels 1 and 2, 2 and 3, and 3 and 4. The attempt to resolve these clashes drives physics forward.

In seeking to resolve these clashes between levels, influences can go in both directions. Thus, given a clash between levels 1 and 2, this may lead to the modification, or replacement of the relevant theory at level 2; but, on the other hand, it may lead to the discovery that the relevant experimental result is not correct for any of a number of possible reasons, and needs to be modified. In general, however, such a clash leads to the rejection of the level 2 theory rather than the level 1 experimental result; the latter are held onto more firmly than the former, in part because experimental results have vastly less empirical content than theories, in part because of our confidence in the results of observation and direct experimental manipulation (especially after expert critical examination and repetition). Again, given a clash between levels 2 and 3, this may lead to the rejection of the relevant level 2 theory (because it is disunified, at odds with the current assumption at level 3); but, on the other hand, it may lead to the rejection of the level 3 assumption and the adoption, instead, of a new assumption (as has happened a number of times in the history of physics, as we have seen). The rejection of the current level 3 assumption is likely to take place if the level 2 theory, which clashes with it, is highly successful empirically, and furthermore has the effect of increasing unity in the totality of fundamental physical theory overall, so that clashes between levels 2 and 4 are decreased. In

145

general, however, clashes between levels 2 and 3 are resolved by the rejection or modification of theories at level 2 rather than the assumption at level 3, in part because of the vastly greater empirical content of level 2 theories, in part because of the empirical fruitfulness of the level 3 assumption (in the sense indicated above).

It is conceivable that the clash between level 2 theories and the level 4 assumption might lead to the revision of the latter rather than the former. This happened when Galileo rejected the then current level 4 assumption of Aristotelianism,[17] and replaced it with the idea that "the book of nature is written in the language of mathematics" (an early precursor of our current level 4 assumption). The whole idea of aim-oriented empiricism is, however, that as we go up the hierarchy of assumptions we are increasingly unlikely to encounter error, and the need for revision. The higher up we go, the more firmly assumptions are upheld, the more resistance there is to modification.

Aim-oriented empiricism is put forward as a framework which makes explicit metaphysical assumptions implicit in the manner in which physical theories are accepted and rejected, and which, at the same time, facilitates the critical assessment and improvement of these assumptions with the improvement of knowledge, criticism being concentrated where it is most needed, low down in the hierarchy. Within a framework of relatively insubstantial, unproblematic and permanent assumptions and methods (high up in the hierarchy), much more substantial, problematic assumptions and associated methods (low down in the hierarchy) can be revised and improved with improving theoretical knowledge. There is something like positive feedback between improving knowledge and improving (low-level) assumptions and methods – that is, knowledge-about-how-to-improve-knowledge. Science adapts its nature, its assumptions and methods, to what it discovers about the nature of the universe. This, I suggest, is the nub of scientific

---

[17] Aristotelianism holds that the earth is at the centre of the universe, objects fall because they seek to get to the centre of the earth, things change because they seek to realize an inherent potential, and heavenly bodies are perfect, unchanging, and move uniformly in circles round the earth.

rationality, and the methodological key to the great success of modern science.

Putting this aim-oriented empiricist methodology into scientific practice gives us our best hope, I claim, of *improving* metaphysical assumptions about the nature of the universe as science proceeds. It provides, what standard empiricism cannot, a rational, if fallible and non-mechanical method for the discovery of new theories, and enables science to improve its methods in the light of what is discovered about the nature of the universe.[18]

Once standard empiricism is recognized to be the untenable view that it is, and aim-oriented empiricism is accepted in its stead, it becomes clear that physicalism – the thesis that the God-of-Cosmic-Power exists – is indeed a basic component of current scientific knowledge. It is more firmly established than any accepted physical theory. We have good grounds, as we have seen, that all accepted theories, despite their great empirical success, are false. Empirically successful theories that clash too severely with physicalism are rejected, for that reason. All scientific revolutions in physics since Galileo have brought physics closer to capturing physicalism as a testable physical theory of all phenomena, in that they have created greater unity in theoretical physics. Physicalism might be overthrown in the future, but it would cause a rupture in science far greater than that brought about by the advent of Einstein's theories, or of quantum theory. Not since Galileo's break with Aristotelianism some 400 years ago have ideas at level 4 been radically revised.

There are two versions of physicalism: reductionist physicalism and experiential physicalism. The first holds that only the physical exists (everything being in principle reducible to fundamental physical entities interacting with one another, $U + V$ of the true, unified physical theory of everything, T. The second holds that the experiential exists *in addition* to the physical, and cannot even in principle be reduced to the physical. In chapter three I spelled out the case for this latter view of experiential physicalism.

---

[18] For these and other benefits of aim-oriented empiricism see works referred to in note 16.

(5) How can *empirical* considerations affect what metaphysical thesis is accepted given that these theses are *metaphysical*, and thus empirically untestable?

The answer to this enigma is I hope clear, as a result of what I said briefly above. Even though a metaphysical thesis is not empirically testable, it may be empirically *fruitful* in the sense that, successive attempts to capture the thesis by means of testable theories – successive testable theories which draw ever closer to the metaphysical thesis – meet with great empirical success. A metaphysical thesis may, in other words, lead to an empirically successful research programme. Given two rival metaphysical theses, M_1 and M_2, M_1 may be much more *empirically fruitful* than M_2, successive attempts to develop theories that accord with M_1 meeting with much greater empirical success than attempts to develop theories in accord with M_2. Aim-oriented empiricism provides the methodological framework best able to help the development of new metaphysical theses likely to be empirically fruitful, and best able to assess the empirical fruitfulness of these metaphysical theses, especially those at level 3.

(6) Does not this view of science, which holds that science rests on *an article of faith*, an *untestable, metaphysical assumption*, amount to a dreadful betrayal of scientific rigour and objectivity?

The answer is a resounding **No!** It is all the other way round. Standard empiricism lacks rigour because it fails to acknowledge highly problematic and influential metaphysical theses implicit in the persistent preference of physics for unified, explanatory theories. Aim-oriented empiricism is more rigorous than standard empiricism because it makes explicit these problematic, influential and implicit assumptions, and subjects them to the best possible critical appraisal, in terms of their empirical fruitfulness.

What is being appealed to here is:

Principle of Intellectual Rigour: An assumption that is substantial, influential, problematic and implicit needs to be made explicit so that it can be critically assessed, so that alternatives can be considered, in the hope that it can be improved.[19]

Standard empiricism violates this principle of intellectual rigour. Aim-oriented empiricism may be regarded as emerging as a result of a series of applications of this principle – the hierarchy of metaphysical theses and associated methods emerging as a result.

(7) What does it *mean* to assert that one metaphysical thesis is an *improvement* over another?

A basic aim of science – or of *natural philosophy*, as aim-oriented empiricist science should perhaps be called – is to discover what kind of universe this is. In pursuing physics we hope to accept metaphysical theses, at levels 3 and 4, which are true. The chances are, however, that at level 3 (and perhaps at level 4 as well) the currently accepted thesis is false. Improving this thesis involves accepting a thesis that is less false, that is closer to the truth, that captures more accurately the nature of that which ultimately determines the way events unfold. But what does it mean to say of two rival metaphysical theses, M_1 and M_2, both false, that M_2 is "closer to the truth" in this way?

We can make sense of this via the notion of one physical theory, T_2, being closer to the truth, T, than another theory, T_1, where T is the as-yet-undiscovered true theory of all physical phenomena. As we have seen, physics advances from one false theory to another, each new theory correcting its predecessor. Thus Newton's theory corrects Kepler's laws. Kepler holds that the planets move in ellipses round the sun. Newton's theory predicts that the planets would move in ellipses if the sun did not move, and the planets did not attract each other. But the sun does move, as a result of being attracted by the planets by gravitation; and furthermore the planets attract each other gravitationally, which makes then deviate from precise ellipses. But even though Newton's and Kepler's theories contradict each other, there is, nevertheless, a sense in which Kepler's theory can be extracted from Newton's. We can do this

---

[19] "Improved" means "made less false". What this means is explicated in the next section; see also Maxwell (2007, chapter 14, section 5, "The Solution to the Problem of Verisimilitude").

in three steps. First, we restrict Newton's theory to systems enclosed in a region of space. Second, we let the masses of all the bodies but one (the most massive) tend to zero. In the limit, the zero-mass bodies trace out paths that are ellipses around the body – the sun – that has retained its mass. The form of Kepler's law has been extracted from Newton's theory. We then reinterpret Kepler's law to apply to solar systems with planets that have non-zero mass. It is this final step of reinterpretation that introduces error (from a Newtonian perspective).

We can now specify what it means to say T_2 is closer to the truth, T, than T_1. This is the case if T_2 can be extracted from T, and T_1 can be extracted from T_2, by means of the kind of steps just indicated in the case of Newton and Kepler, but not *vice versa*. That is, T_2 cannot be extracted, in this way, from T_1.

Consider, now, the false metaphysical theses M_1 and M_2. M_2 is closer to the truth, T, than M_1 is if there is a theory, T_2, compatible with M_2, and a theory T_1 compatible with M_1 such that T_2 is closer to the truth, T, than T_1 is, but not *vice versa*. (We require, that is, that there is no T_1 compatible with M_1 and no T_2 compatible with M_2 such that T_1 is closer to the truth, T, than T_2 is.)[20]

(8) How is it be possible for what exists at one instant to determine what exists next, given David Hume's decisive arguments which show that this is not possible?

The God-of-Cosmic-Power is, in one crucial respect, quite different from God as conceived by Christianity or Islam. It is impersonal, a physical entity, utterly indifferent to and unaware of the suffering of humanity that it causes, and incapable of knowing anything about it (or anything else). It has no consciousness, no knowledge, no purpose. Because it is an impersonal physical entity incapable of knowing about human suffering, or indeed anything, we can forgive It all the terrible things It does.

In other respects, however, the God-of-Cosmic-Power does have some of the attributes of the God of traditional Christianity or Islam.

---

[20] This solves a long-standing problem in the philosophy of science known as the problem of "verisimilitude". For further details see my 2007a, *From Knowledge to Wisdom*, chapter 14, section 5).

It is omnipresent, eternal and unchanging. And It is omnipotent. It determines the way all events unfold (perhaps probabilistically). Given the existence of **U**, the God-of-Cosmic-Power at some instant in time, and given **V**, the instantaneous variable state of the universe at that moment, subsequent states are necessarily determined (perhaps probabilistically).

But how can this be? A famous argument due to David Hume is generally held to have demolished this idea that *what exists* at one instant necessarily determines what exists at subsequent instants. Hume argued that we can always imagine a change in the course of nature, a change in physical laws, which proves that such a change is always possible.[21]

As it happens, this argument of Hume was decisively demolished some forty years ago (at the time of writing) in my second published paper.[22] The counter argument to Hume goes like this. Theoretical physics needs to be reinterpreted so that theories are understood to attribute *necessitating properties* to postulated physical entities. A necessitating physical property is somewhat like an ordinary, common sense dispositional property, such as solid, rigid, sticky, opaque or inflammable. In every case, in attributing such a property to an object we imply something about how that object would behave were it to be subjected to such and such conditions. Thus, in declaring an object to be inflammable, we imply that it will burst into flames if exposed to a naked flame. If it does not, then it is not inflammable. In other words, attributing such a common sense dispositional property to an object implies that that object will exhibit law-like behaviour in such and such conditions.

The same goes, but with far greater precision and content, for necessitating physical properties of physical entities. To say of a particle that it has an electric charge is to say that it will accelerate in such and such a law-like way in an electric field. *If* it is electrically charged, and *if* it is in the electric field then, *of necessity*, it will behave in the prescribed law-like way. If it doesn't, then either the

---

[21] "We can at least conceive a change in the course of nature; which sufficiently proves that such a change is not absolutely impossible", Hume (1959), p. 91. See also Part III, sections II-VI and XIV-XV.

[22] See my (1968a). See also my (1998), pp. 141-155.

particle wasn't charged, or there wasn't the electric field. But if both exist then, of necessity, the particle moves in the prescribed law-like fashion. The necessitating property of what may be called "classical electric charge" is such that the laws of classical electrodynamics are implicit in the meaning of the term used to attribute this property to a particle. As a result, descriptive propositions which do no more than specify the instantaneous state of an isolated physical system can imply subsequent states of the system. The descriptive propositions, in attributing necessitating properties (such as "classical electric charge") to physical entities of the system, implicitly specify the laws these entities must obey, and it is this which enables the descriptive propositions to imply propositions which describe subsequent states of the system.[23]

It is in this way that what exists at one instant can determine what exists at subsequent instants. Even if such "necessary connections" exist between successive states of affairs, it doesn't mean we can know with certainty that they exist. In order to know that, we would have to know with certainty that the relevant physical entities exist, and have the relevant necessitating properties, and that we can never know with certainty, any more than we can know with certainty that physical theories are true.

However, if a physical theory, such as classical electrodynamics, is to attribute such necessitating properties to particles and fields it must be formulated in a way which, in one respect, differs radically from the way such a theory would ordinarily be formulated. The laws of the theory must all be regarded as *analytic* propositions. That is, they must be interpreted to be statements like "all bachelors are unmarried", true in virtue of the meaning of the constituent terms, and devoid of factual content. If it is built into the *meaning* of "classical electric charge" that a particle with this property obeys the laws of classical electromagnetism, then a law asserting that a particle with this property obeys the laws of classical electrodynamics tells us nothing new. It *must* of course be true. It is

---

[23] Strictly speaking, it is only a specification of what exists at an instant throughout the universe that can logically imply subsequent states – unless, in the case of an isolated system, one adds the assertion "The system will remain isolated for the period in question".

true in virtue of the meaning that has been assigned to "classical electric charge". Such a law tells us nothing empirical; it just spells out explicitly what is implicit in the meaning of "classical electric charge".

A physical theory that can be used to attribute necessitating properties to physical entities such as particles must, then, be interpreted in such a way that all the laws of the theory merely make explicit what is implicit in the meaning of constituent terms. All the laws of such a theory are true *analytically*, in virtue of the meaning of the constituent terms, and do not assert anything empirical or factual.

How then, it will be asked, can the theory make any kind of factual, empirical assertion? The answer is that the empirical content of the theory is contained entirely in the *factual* assertion that such and such physical entities do actually exist with such and such necessitating properties. Classical electrodynamics interpreted in this "conjecturally essentialistic" way is an empirical theory because, although all the laws are analytic and without factual content, the theory asserts that particles exist that have the necessitating property of classical electric charge, and this assertion (when formulated in a bit more detail) is empirical in character, and capable of being falsified. When a theory, interpreted in this essentialistic way, is falsified empirically, and replaced by a better essentialistic theory, the laws of the first theory are not refuted. Rather, what happens is that the necessitating properties attributed to physical entities by the first theory are found not to exist. The new theory attributes new necessitating properties to physical entities it postulates.

What becomes of Hume's argument that we can always imagine a change in the course of nature, and hence there cannot be necessary connections between successive states of the universe? We can certainly imagine that phenomena, that appear to be like the phenomena that surround us, might suddenly start behaving very differently. But that does not establish that it is, in reality, possible. For, it may be that the phenomena surrounding us do really possess physical necessitating properties of the kind I have just indicated. If they do, and they are of the right kind, then it is logically impossible for phenomena around us abruptly to start behaving differently. Hume's argument collapses.

153

What is involved here, as I have indicated above, is a change in the way physics is interpreted and understood. At least since Newton, the tendency has been to interpret physics as specifying no more than regularities in nature, or "the laws of nature". Even today, physicists speak of "the rules" of quantum theory, or classical electrodynamics. But this traditional way of understanding physics is the outcome of failing to appreciate what the implications are of cutting God in half.

If the traditional Christian God exists, it makes perfect sense to restrict physics to discovering the rules or laws observed by natural phenomena. For in this case, it is ultimately God who makes phenomena conform to the rules or laws, and it would hardly be possible for physics to discover how God does this. Physics can *describe* what goes on, but cannot *explain*, in the sense of depict what it is that makes phenomena conform to the rules, for this is God, and God, presumably, lies beyond the scope of physics.

However, cut God in half, and this traditional conception of physics leaves us with a profound mystery. What is it, in existence, that *makes* physical phenomena conform to the rules? Having got rid of the traditional Christian God to perform this task, some kind of substitute must be found. Otherwise one has the utterly unintelligible state of affairs that phenomena occur in accordance with precise rules or laws but nothing exists to ensure that this happens. That the rules continue to be observed would be nothing short of an absolute miracle. As physical theories are developed which specify rules of ever wider and wider scope, physics would succeed, not in making phenomena more and more comprehensible but, quite the contrary, more and more *incomprehensible.*

The solution is to reinterpret physics in the way indicated above so that it seeks to specify the physical nature of *what it is in existence* which ensures, which necessarily determines, the way events unfold. The true unified physical "theory of everything", if ever we discover it, will attribute necessitating properties to U; it will thus specify the physical nature of the God-of-Cosmic-Power. A precise specification of $U + V$, at some instant, logically implies specifications of the states of $U + V$ at subsequent instants. Genuine, fully fledged physical explanations as to why phenomena occur as they do become possible, even though the theories we employ at present only depict what really makes events occur as they do in approximate, limited, partial ways.

(9) Does "The God-of-Cosmic-Power exists" really amount to a meaningful, substantial assertion?

We can get a hint of how substantial a proposition is by considering what it *denies* – what it is incompatible with. The more a proposition denies, the more it asserts. (If nothing is denied, nothing is asserted.) Consider "the God-of-Cosmic-Power exists" interpreted to mean merely that the true physical "theory of everything" is unified to degree 500 (in a type 1 to 5 way). Even this relatively open-ended, insubstantial version of the assertion still makes a very substantial assertion about the nature of the universe. It asserts that nature is *physical* in character. It denies that God, gods, devils, ghosts, poltergeists, Cartesian minds or other non-physical entities exist and interact with anything physical. It denies that the universe is such that a physical theory disunified to degree greater than 500 is required to depict the way phenomena occur. If "the God-of-Cosmic-Power exists" is interpreted in a much more restrictive way to assert that the universe is such that the true physical "theory of everything" is perfectly unified, with $N = 1$, then the assertion becomes all the more substantial. It denies that any physical theory can be true which is not true when interpreted to be about all phenomena, or which is to any extent disunified.[24] It denies, in other words, that any current theory of physics is true.

(10) Does not Quantum Theory Refute the Thesis that only Unified Theories are Accepted in Physics?

The argument of this chapter depends crucially on the thesis that, in physics, only unified theories are accepted. But quantum theory, given its orthodox interpretation, goes against this thesis. Orthodox quantum theory (OQT) is about the results of performing

---

[24] One needs to add that only *precise* disunified theories are being considered. Given the true unified theory of everything, T, endlessly many *imprecise* disunified theories could be concocted which are implied by T, and which are thus true.

measurements on quantum systems – electrons, photons, atoms, etc. It is not about electrons, photons, atoms *per se*. This means OQT must include some part of classical physics for a treatment of measurement. If OQT is applied to the measuring apparatus in addition to the quantum system being measured, no definite prediction is forthcoming. A second measuring device would be required, to be described by means of some part of classical physics, to measure the first instrument, for a definite experimental result to emerge. This means that the theory that makes physical predictions consists of quantum postulates *plus* postulates of some part of classical physics for a treatment of measurement. And this in turn means that OQT is a profoundly disunified theory, in that it consists of two incompatible parts, a quantum part applicable to the quantum system, and a classical part applicable to the measuring instrument.

OQT is severely disunified in this way because of the failure to solve the quantum wave/particle problem. Quantum entities – electrons, photons, even atoms – seem to be both particle-like and wave-like. For example, send an electron through a screen with two slits in it so that it strikes a photographic plate beyond, and the electron will be recorded as a dot on the photographic plate which, along with many other experiments, seems to indicate that the electron is a particle. Send many such electrons through the two-slitted screen, all with the same velocity, one at a time, and the photographic plate records each electrons as a minute dot. But all the dots together on the photographic plate add up to an interference pattern – dark bands where many electrons have arrived, interwoven with light bands where few have arrived. This interference pattern of dots can be explained if each electron is assumed to be a wave-like entity when it encounters the two-slitted screen. This wave-like entity passes through both slits, and travels towards the photographic plate. In some regions of the plate, where a crest of the wave arrives from one slit, a trough arrives from the other slit. The two cancel each other out, there is no wave, and a light band results. At other regions of the plate, where a crest arrives from one slit, a crest arrives from the other slit as well. The two support each other, waves are high, and a dark band results. OQT makes all this precise. It specifies precisely the wave-like character of the electrons in flight, and predicts that high waves mean that there is a high probability that the electron will be detected in that region, and low waves mean a low probability of detection.

OQT successfully predicts the results of the experiment, but only at the cost of generating a dreadful paradox. The electron is, it seems, a wave-like entity when it passes through the two-slitted screen and travels towards the photographic plate, but a particle-like entity when it hits the plate. How can an electron be both wave-like (spread continuously throughout a region of space) and particle-like (concentrated in a tiny region of space) at one and the same time? This is the quantum wave/particle paradox. The authors of OQT, Heisenberg, Bohr, Born and others, decided the problem could not be solved. They *evaded* it by developing OQT as a theory about the results of performing *measurements* on quantum systems. As a result, OQT did not have to say what a quantum system is – wave, particle, or something else – when not being measured.

The drawback in this, however, is the one that we have seen. OQT is a severely disunified theory.

It might be argued that *all* physical theories that make predictions must be disunified in a similar way. Thus, Newtonian theory, employed to predict the position of a planet in the sky, would need to call upon additional theory about light and telescopes to predict what is observed with a telescope. The big difference between Newtonian theory and OQT, however, is that Newtonian theory has physical content, and can issue in physical predictions, without calling upon additional theory, whereas OQT does not, and cannot. Newtonian theory predicts that the planet has a certain position at a certain time, whether anyone observes it or not.[25] OQT can make no such physical prediction about the unobserved electron, precisely because no solution to the wave/particle problem is forthcoming.

---

[25] Ignore, here, the irrelevant point that additional theory is needed to predict that the planet holds its shape, and does not implode under its own gravitational attraction. Strictly speaking, we need to consider Newtonian theory applied to classical point-particles with mass.

Despite this, OQT has been widely accepted by the physics community since its discovery in 1925 and 1926 by Heisenberg, Schrödinger and Born. Does not this decisively refute the thesis that physicists only accept *unified* theories, and thus refute the entire argument of this chapter?

Not at all. What the long-standing and general acceptance of the severely disunified theory of OQT shows is just how important the argument of this chapter is for physics itself. OQT was regarded as acceptable by a majority of physicists for decades after its discovery because during this time the entire scientific community took for granted some version of standard empiricism. Given this state of affairs, it is not at all surprising that a majority of physicists should have accepted OQT. For standard empiricism holds that the decisive factor in deciding what theory should be accepted or rejected is empirical success. Considerations of unity and simplicity play a role as well, but not one which over-rides empirical considerations. And in any case standard empiricism leaves obscure what the unity or simplicity of a theory is. This further weakens the role that these considerations play in deciding what theory is to be accepted or rejected – as long as standard empiricism is accepted. One very striking feature of OQT is its astonishing empirical success. No other theory in physics successfully predicts such a wealth of diverse phenomena – without, apparently, running into serious experimental difficulties anywhere. Thus, given standard empiricism, there seem to be overwhelming grounds for accepting OQT. In these circumstances, it is not at all surprising that OQT was widely accepted by the physics community not long after it was put forward.

But if, from the 1920s onwards, aim-oriented empiricism had been accepted by most physicists, OQT would have been regarded as highly problematic, and in urgent need of improvement. For, despite its great empirical success, OQT fails to satisfy the all-important requirement of aim-oriented empiricism of unity. OQT would have been accepted as the best predictive scheme available in lieu of anything better. It would have been appreciated that, in order to develop an acceptable theory of the quantum domain, the quantum wave/particle problem has to be solved, so that quantum theory can have its own quantum ontology and can be interpreted as being about quantum systems *per se*, measurement being removed from the basic postulates of the theory altogether. Solving the quantum wave/particle dilemma would have become a major concern of theoretical physics – something which did not actually happen, given general acceptance of standard empiricism.

It is noteworthy that a few of the great theoretical physicists associated with the creation of quantum theory did take the "aim-oriented empiricist" attitude towards OQT just described. These include Einstein, Schrödinger, de Broglie and von Laue. Einstein remarked in a letter to Schrödinger "You are the only contemporary physicist, besides Laue, who sees that one cannot get around the assumption of reality – if only one is honest. Most of them simply do not see what sort of risky game they are playing with reality – reality as something independent of what is experimentally established" (Einstein, 1986, p. 39). It should be noted that Einstein, after his discovery of general relativity in 1915, held a view close to aim-oriented empiricism, as I have argued elsewhere.[26]

Thus, whether OQT is held to be unproblematically acceptable or not depends crucially on which of standard or aim-oriented empiricism is accepted. Furthermore, whether the wave/particle problem is seen as a fundamental problem of theoretical physics or not depends on which of standard or aim-oriented empiricism is accepted. All of which highlights the importance of the issue for theoretical physics.

To all this it may be objected: But the historical fact that OQT was widely accepted refutes the central argument for aim-oriented empiricism. For that argument depends on the claim that physicists only ever accept *unified* theories. That claim is plainly false. OQT is a theory that was widely accepted despite being very seriously *disunified*.

My answer is: No. In the first place, OQT is a special case. It was not appreciated for decades after its initial discovery – if it is appreciated even today – that OQT is unacceptably disunified.

---

[26] See my (1993a), pp. 275-305.

This is something I sought to point out in the 1970s and later.[27] Secondly, if the special case of OQT became the general rule in physics, and empirically successful theories as disunified as OQT became acceptable quite generally, this would spell the end of progress in theoretical physics. For the argument spelled out above is quite general. Given any accepted unified theory, endlessly many empirically more successful disunified rivals can always be concocted. If empirical success became the over-riding factor in deciding what theories should be accepted and rejected, as standard empiricism holds, then physicists would be obliged persistently to choose these empirically successful disunified rivals – disastrous choices for progress in physics, as every physicist would acknowledge. Physics makes progress because physicists in practice, but somewhat surreptitiously, implement a view close to aim-oriented empiricism (in excluding empirically successful disunified rival theories from consideration), while all the time proclaiming standard empiricism. As Einstein again acutely said "If you want to find out anything from the theoretical physicists about the methods they use…don't listen to their words, fix your attention on their deeds" (Einstein, 1973, p. 270).

To sum up the argument of this section so far, there are decisive grounds for rejecting standard empiricism and accepting aim-oriented empiricism instead; furthermore, given the latter, OQT is at best a defective theory, despite its great empirical success, because of its disunity.

But are there grounds for holding OQT to be defective that are quite independent of whether one accepts standard or aim-oriented empiricism? Would the judgement of a physicist in 1930, who accepted aim-oriented empiricism, that OQT is defective, have been correct?

---

[27] In a series of papers I pointed out, again and again, that bringing measurement into the basic postulates of the theory – as a result of the failure to solve the quantum wave/particle problem – means that OQT is a severely *ad hoc* or non-explanatory theory, being made up of two incompatible parts, a quantum part and a classical part (for measurement): see Maxwell (1972, 1973, 1976b, 1982, 1988, 1993b, 1994, 1995, 1998 ch. 7, 2004b, 2010b).

There are indeed such grounds. The failure of OQT to solve the wave/particle problem – and the resulting need for the theory to be formulated as being about the results of performing measurements on quantum systems (instead of being about quantum systems *per se*) – results in the theory having a number of defects in addition to that of disunity.[28] It means the theory lacks *precision*, it being impossible to specify precisely the physical conditions for measurement. It means the theory lacks *explanatory power*. The theory ought to be able to predict and explain the approximate success of classical physics, but this OQT cannot do entirely successfully since it must presuppose some part of classical physics to describe measuring instruments, as a matter of conceptual necessity, and not just practical convenience. It means OQT resists *unification with general relativity*, since such a unification, granted it inherits the measurement-dependence of OQT, would have to be a theory about performing measurements on space-time with measuring instruments placed outside space-time, a difficult measurement to perform.[29] For similar reasons, OQT *cannot be applied to the entire cosmos*, relevant especially in considering states of the universe soon after the big bang. Such an application would require a measuring instrument to be placed outside the cosmos! Most serious of all, perhaps, OQT fails to answer two basic questions about the nature of the quantum domain, namely: Is the quantum domain fundamentally deterministic or probabilistic? What sort of physical entities are electrons, photos, atoms? OQT seems to imply that probabilistic events occur when measurements are made, but the theory is ambivalent as to whether probabilistic events really do occur in nature, or whether they only seem to occur, when we intervene and

---

[28] It is sometimes argued that quantum field theory solves the wave/particle problem. This is not the case at all. Quantum field theory is just as dependent on measurement for its physical interpretation as non-relativistic OQT is.

[29] General relativity is a theory about space-time and not, in the first instance, a theory about performing measurements on space-time. There is thus a basic structural difference between general relativity and OQT which, along with other problems of course, creates a difficulty for unification.

make a measurement. And, of course, as long as no solution to the wave/particle is forthcoming, OQT fails to say what sort of physical entities electrons, photons and atoms are.[30]

When I first began to work on quantum theory around 1970, OQT was so firmly, or even dogmatically, accepted that calling it into question could threaten a physicist's career. Almost all physicists regarded those few who did question OQT a lunatic fringe. Since then, the situation has changed dramatically. All physicists are aware of at least some of the problems OQT faces, and many find OQT unsatisfactory. Rival interpretations and versions of quantum theory have been developed. There is the de Broglie-Bohm interpretation (quantum entities are particles with an associated wave-like quantum potential). There is the many-worlds or Everett interpretation (according to which, whenever a measurement-type interaction occurs with N possible outcomes, all these occur but the other N-1 worlds become undetectable to us in our world). There is the stochastic theory of Ghirardi, Rimini and Weber (according to which quantum systems spontaneously and probabilistically localize). And there is decoherence theory (according to which it is the environment, not measurement, which induces ostensible probabilistic events and "the reduction of the wave packet").[31] This does not, however, eliminate the disunity of OQT, nor some of the other defects.

My own view is that quantum theory is trying to tell us that Nature is fundamentally probabilistic. If electrons, atoms and other quantum entities interact with one another *probabilistically*, it follows at once that they will be very different from any physical entities of *deterministic* classical physics. "Is the electron a (classical) wave or a particle?" is thus the *wrong* question. The right question is: "What kind of unproblematic, fundamentally probabilistic physical entities are there, as possibilities, and can we see quantum entities as a specific variety of one of these unproblematic possibilities? I think we can. Elsewhere, I have

---

[30] I have spelled out these defects of OQT at greater length elsewhere: see works referred to in note 27. See also Bell (1987).

[31] For an excellent survey of diverse interpretations and versions of quantum theory, see Wallace (2008).

suggested that electrons, atoms, etc. can be construed as fundamentally probabilistic entities whose deterministic, physical evolution through space in time is specified by Schrödinger's time-dependent wave equation, but whose intermittent, abrupt, probabilistic changes of physical state are specified by a modified version of Born's postulate.[32] Macroscopic objects are the outcome of these two kinds of transitions occurring in rapid succession. The key problem confronting this "popensiton" view is to specify the precise quantum theoretic conditions for probabilistic transitions to occur, no mention being made of measurement, or some surrogate. My suggestion, here, is that probabilistic transitions occur, roughly, when new particles, or bound systems, are created. I argue that this fully micro-realistic, fundamentally probabilistic version of quantum theory reproduces all the empirical success of OQT, but nevertheless is also empirically distinct from OQT for as yet unperformed experiments.[33]

In my view, if aim-oriented empiricism had been the orthodox view in the 1920s, it would have been dazzlingly clear to everyone *that* OQT is defective, *why* it is defective, and *what* needs to be done, in general terms, to remove the defects: namely, solve the wave/particle problem. If this had been generally understood in the 1920s, we might now have solved the problems of quantum theory which still remain, at the time of writing, over 80 years later, with no agreed solution.

The long-standing acceptance of the disunified theory of quantum theory does not refute the central argument of this chapter. Indeed, as I have just said, if that argument had been understood by physicists in the 1920s, we might now possess a genuinely acceptable, unified version of quantum theory accepted by everyone – something we still do not have.

---

[32] Born's postulate, a key ingredient of OQT, specifies what happens, in general probabilistically, when measurements are made.
[33] For details see my (1988, 1994, 2004b, 2010b).

CHAPTER SIX

WISDOM-INQUIRY

What kind of inquiry can best help humanity sustain and create what is genuinely of value in life in the physical universe? What kind of inquiry can best help us make progress towards as good, civilized, and enlightened a world as possible? Or, in other words, what kind of inquiry best helps us promote the flourishing of the God-of-Cosmic-Value, here on earth, within the God-of-Cosmic-Power?

The answer is wisdom-inquiry. I gave a preliminary sketch of what wisdom-inquiry is, and why it is needed, towards the end of chapter two. Here, I spell out a second argument in support of wisdom-inquiry. This builds on, and further develops, the picture of wisdom-inquiry I have already given.

Two Great Problems of Learning

Two great problems of learning confront humanity: learning about the universe and ourselves as a part of the universe; and learning how to become civilized.[1]

We solved the first problem in the 17th century, with the creation of modern science. A *method* was discovered for progressively improving knowledge and understanding of the natural world, the famous empirical method of science. There is of course much that we still do not know and understand, three or four centuries after the birth of modern science; nevertheless, during this time, science has immensely increased our knowledge and understanding, at an ever accelerating rate. And with this unprecedented increase in scientific knowledge and understanding has come a cascade of technological discoveries and developments which have transformed the human condition.[2]

_____

[1] This nomenclature is a bit misleading. We do not have two distinct problems of learning here, as the second problem includes the first. We should construe the task of creating civilization as including the task of improving knowledge.

But the second great problem of learning has not yet been solved. And this puts us in a situation of unprecedented peril. Solving the first problem without also solving the second is bound to create a situation of great danger. Indeed, all our current global problems can be traced back, in one way or another, to this source.

With rapidly increasing scientific knowledge comes rapidly increasing technological know-how, which brings with it an immense increase in the power to act. In the absence of a solution to the second great problem of learning, the increase in the power to act may have all sorts of good consequences, but will as often as not have all sorts of harmful consequences as well, whether intended or not.

Just this is an all too apparent feature of our world. Science and technology have been used in endless ways for human benefit, but have also been used to wreak havoc, whether intentionally, in war, or unintentionally (initially at least), in long-term environmental damage – a consequence of growth of population, industry and agriculture, made possible by growth of technology. Global warming, rapid extinction of species and destruction of natural habitats such as tropical rain forests, depletion of natural resources, the threats posed by modern armaments whether conventional, chemical, biological or nuclear, vast inequalities of wealth and power across the globe, even the aids epidemic (aids being spread by modern travel): these have all been made possible by modern science and technology. As long as humanity's power to act was limited, lack of wisdom, of enlightenment did not matter too much: humanity lacked the means to inflict too much damage on the planet.[3] But with the immense increase in our powers to act that

---

[2] There is a long-standing debate as to whether technology emerges from science, or develops independently, or actually contributes to science (as in the case of the steam engine leading to the development of thermodynamics). I sidestep this debate, here, and assume, merely that, as far as modern science is concerned, science and technology developed in tandem with each other, each contributing to the development of the other.

[3] Humans have been causing some environmental damage for centuries. Aldous Huxley cites the ancient destruction of the ceders of Lebanon as an example; see Huxley (1980), pp. 21-22. For a discussion of the role

we have achieved in the last century or so, our powers to destroy have become unprecedented and terrifying: global wisdom has become, not a luxury, but a necessity. All our distinctively 21st century global problems, to repeat, have arisen because we have solved the first great problem of learning without also solving the second problem. Solving the first great problem of learning without solving the second is bound to put humanity into a situation of great danger, *and has in fact done just that.*

It has become a matter of extreme urgency, for the future of humanity, that we discover how to solve the second great problem of learning, not so that we achieve instant global wisdom, but so that we learn how to make gradual social progress towards a wiser world.[4]

Many blame science for our problems. There is a sense in which they are right to do so. Science is, in an important sense, an underlying cause of all our current global problems, as I shall show in chapter nine. Blaming science must not, however, blind us to the possibility that science contains a vital key to solving the second great problem of learning. The important point to appreciate is that our solution to the first great problem of learning has been astonishingly successful. It is just this immense intellectual success of science in improving knowledge which has made possible *both* a multitude of benefits to humanity *and* a great deal of harm.

The crucial question arises: Can we learn from our solution to the first great problem of learning how to solve the second problem? Can the progress-achieving methods of science be generalized so that they become fruitfully applicable to the

---

of early man in causing extinction of species see Holdgate (1996), pp. 1-10.

[4] We need to solve *both* great problems of learning in order to solve our *fundamental* problem: How can we best help the God-of-Cosmic-Value to flourish in the God-of-Cosmic-Power? It is clearly not enough to solve the first problem – the problem of learning about the universe – the God-of-Cosmic-Power. Wisdom-inquiry – I argue – is the solution to *both* great problems of learning, and thus the solution to our fundamental religious problem of learning.

immense task of making social progress towards a more civilized world?

The Enlightenment

The idea of learning from the solution to the first great problem of learning how to solve the second problem is an old one. It goes back to the Enlightenment of the 18th century. Indeed, this was the basic idea of the *philosophes* of the Enlightenment – Voltaire, Diderot, Condorcet et al.: to learn from scientific progress how to achieve social progress towards world enlightenment.

The best of the *philosophes* did what they could to put this immensely important idea into practice, in their lives. They fought dictatorial power, superstition, and injustice with weapons no more lethal than those of argument and wit. They gave their support to the virtues of tolerance, openness to doubt, readiness to learn from criticism and from experience. Courageously and energetically they laboured to promote rationality in personal and social life.[5]

Unfortunately – as we saw in chapter two – in developing the Enlightenment idea intellectually, the *philosophes* blundered. They developed the Enlightenment programme in a seriously defective form, and it is this immensely influential, defective version of the programme, inherited from the 18th century, which may be called the "traditional" Enlightenment, that is built into academic inquiry as carried on in universities today. Our current traditions and institutions of learning, when judged from the standpoint of helping us learn how to become more enlightened, are defective and irrational in a wholesale and structural way, and it is this which, in the long term, sabotages our efforts to create a more civilized world, and prevents us from avoiding the kind of horrors we have been exposed to during the 20th century and the first years of the 21st century – wars, third-world poverty, environmental degradation.

The *philosophes* of the 18th century assumed, understandably enough, that the proper way to implement the Enlightenment programme was to develop social science alongside natural

---

[5] As I mentioned in chapter two, the best overall account of the Enlightenment that I know of is still Gay (1973).

science. Francis Bacon had already stressed the importance of improving knowledge of the natural world in order to achieve social progress.[6] The *philosophes* generalized this, holding that it is just as important to improve knowledge of the social world. Thus the *philosophes* set about creating the social sciences: history, anthropology, political economy, psychology, sociology.

This had an immense impact. Throughout the 19th century the diverse social sciences were developed, often by non-academics, in accordance with the Enlightenment idea.[7] Gradually, universities took notice of these developments until, by the mid 20th century, all the diverse branches of the social sciences, as conceived of by the Enlightenment, were built into the institutional structure of universities as recognized academic disciplines.

But, from the standpoint of creating a kind of inquiry designed to help humanity learn how to become civilized, all this amounts to a series of monumental blunders. So severe are these blunders that many today do not even perceive academia as having, as its basic goal, to help humanity create a better world.

In order to implement properly the basic Enlightenment idea of learning from scientific progress how to achieve social progress towards a civilized world, it is essential to get the following three things right.

1. The progress-achieving methods of science need to be correctly identified.

2. These methods need to be correctly generalized so that they become fruitfully applicable to any worthwhile, problematic human endeavour, whatever the aims may be, and not just applicable to the endeavour of improving knowledge.

3. The correctly generalized progress-achieving methods then need to be exploited correctly in the great human endeavour of trying to make social progress towards an enlightened, civilized world.

---

[6] For the importance of Francis Bacon for the Enlightenment see: Gay, op. cit., vol. 1, pp. 11-12 and p. 322.
[7] Mill, Marx, Durkheim and Weber were all in thrall to the traditional Enlightenment.

Unfortunately, the *philosophes* of the Enlightenment got all three points disastrously wrong. They failed to capture correctly the progress-achieving methods of natural science; they failed to generalize these methods properly; and, most disastrously of all, they failed to apply them properly so that humanity might learn how to become civilized by rational means. That the *philosophes* made these blunders in the 18th century is forgivable; what is unforgivable is that these blunders still remain unrecognized and uncorrected today, over two centuries later. Instead of correcting the blunders, we have allowed our institutions of learning to be shaped by them as they have developed throughout the 19th and 20th centuries, so that now the blunders are an all-pervasive feature of our world. Recent developments in Universities, indicated in chapter 2 (see notes 3 and 18), have not yet gone nearly far enough to correct the blunders we have inherited from the *philosophes*.

The Enlightenment, and what it led to, has long been criticized, by the Romantic movement, by what Isaiah Berlin has called 'the counter-Enlightenment', and more recently by the Frankfurt school, by postmodernists and others.[8] My objection to the traditional Enlightenment is very different. In particular, it is the very opposite of all those anti-rationalist, romantic and postmodernist criticisms which object to the way the Enlightenment gives far too great an importance to natural science and to scientific rationality. What is wrong with the traditional Enlightenment, and the kind of academic inquiry we now possess derived from it, is not too much 'scientific rationality' but, on the contrary, not enough. It is the glaring, wholesale *irrationality* of contemporary academic inquiry, when judged from the standpoint of helping humanity learn how to become more civilized, that is the problem.

---

[8] See Berlin (1999); and Berlin (1980), pp. 1-24. For a clearly written, recent, sympathetic but critical discussion of criticisms of the Enlightenment, from Horkheimer and Adorno, via Lyotard, Foucault, Habermas and Derrida to MacIntyre and Rorty, see Gascardi (1999). For less sympathetic criticisms of postmodernists' anti-rationalism see: Sokal and Bricmont (1998); Gross et al. (1996); and Koertge (1998).

The "New" Enlightenment

What exactly are the three blunders of the traditional Enlightenment, as embodied in academic inquiry today, and what needs to be done to put them right? Let us take the three blunders in turn.

The *first* blunder was dealt with in the last chapter. It concerns the nature of the progress-achieving methods of science. By and large, scientists and philosophers of science today make the assumption, inherited from the Enlightenment,[9] that the basic aim of science is knowledge of truth, the basic method being to assess theories impartially with respect to evidence alone, *no permanent assumption being made about the nature of the universe independent of evidence.* Choice of theory in science may be influenced by such considerations as the relative simplicity, unity or explanatory power of the theories in question, in addition to empirical considerations; this is permissible, as long as it does not involve assuming, permanently, that nature herself is simple, unified or comprehensible.[10]

But, as we saw in the last chapter, this orthodox *standard empiricist* view concerning the aim and methods of science is

---

[9] The *philosophes* of the Enlightenment tended to assume that the triumph of Newtonian science over Cartesian science meant also the triumph of Newtonian inductivist methodology over Cartesian rationalism. They tended to espouse the extreme empiricism of Bacon and Locke, rejecting the rationalism of Descartes. But it is perhaps over simplistic to interpret all the *philosophes* of upholding one or other version of standard empiricism. Kant hardly fits into such a picture. More to the point, d'Alembert asserted that "The universe, if we may be permitted to say so, would only be one fact and one great truth for whoever knew how to embrace it from a single point of view"; D'Alembert (1963), p. 29. This is perhaps compatible with, but hardly conforms to the spirit of standard empiricism.

[10] A classic statement of this widely held view is given by Karl Popper: "... in science, only observation and experiment may decide upon the *acceptance or rejection* of scientific statements, including laws and theories"; Popper (1963), p. 54. For an indication of just how widely held standard empiricism is, see my (1998), pp. 38-45.

170

untenable.[11]   Given any unified, accepted theory of physics, however well verified empirically, there will always be infinitely many *disunified* rival theories, equally well or even better supported by the evidence, which make different predictions, in an arbitrary way, for phenomena not yet observed.   In persistently rejecting these disunified, empirically more successful rivals, physics thereby implicitly makes a substantial, highly problematic, metaphysical (i.e. untestable) assumption about the nature of the universe, to the effect that it is such that all such disunified theories are false.   Precisely because this assumption that the universe is more or less physically comprehensible is substantial, influential and highly problematic, it needs to be made explicit within physics so that it can be critically assessed and improved, as physics proceeds.

Standard empiricism seriously misrepresents the basic intellectual aim of science – of theoretical physics in particular. The real aim is to improve knowledge, not of truth *as such*, but rather truth *presupposed to be more or less physically comprehensible or explanatory in some way*.   The empirical method of science cannot get off the ground unless some substantial, but highly problematic assumption is made about the universe, to the effect that it is such that explanations for phenomena exist to be discovered – the universe being *comprehensible* in some way, to some extent at least.   As we saw

---

[11] It is worth noting that Newton upheld a conception of natural philosophy (natural science) that is, in important respects, more sophisticated than standard empiricism, presupposed by so many 20th century scientists and philosophers of science. Newton formulates three of his four rules of reasoning in such a way that it is clear that these rules make assumptions about the nature of the universe. Thus rule 1 asserts: "*We are to admit no more causes of natural things than such as are both true and sufficient to explain their appearances.*" And Newton adds: "To this purpose the philosophers say that nature does nothing in vain, and more is in vain when less will serve; for Nature is pleased with simplicity, and affects not the pomp of superfluous causes." See Newton (1962), vol. 2, p. 398. Newton understood that persistently preferring simple theories means that Nature herself is being persistently assumed to be simple (which violates standard empiricism).

in the last chapter, we need to adopt and put into scientific practice a new *aim-oriented empiricist* conception of science which construes science as making a hierarchy of assumptions concerning the comprehensibility and knowability of the universe.

This *aim-oriented empiricist* methodology, in stark contrast to current orthodoxy, is the key to the success of modern science. The basic aim of science of discovering how, and to what extent, the universe is comprehensible is deeply problematic; it is essential that we try to improve the aim, and associated methods, as we proceed, in the light of apparent scientific success and failure. In order to do this in the best possible way we need to represent our aim at a number of levels, from the specific and problematic to the highly unspecific and unproblematic, thus creating a framework of fixed aims and meta-methods within which the (more or less specific) aims and methods of science may be progressively improved in the light of apparent empirical success and failure, as depicted in diagram 4 of Chapter five. The result is that, as we improve our knowledge about the world, we are able to improve our knowledge about how to improve knowledge, the methodological key to the rapid progress of modern science. There is something like positive feedback between improving scientific knowledge, and improving knowledge-about-how-to-improve-knowledge. All this has long gone on in scientific practice – science would not have achieved its success if it had not – but awareness of it has been obscured by the scientific community's allegiance to standard empiricism.[12]

Adoption and explicit implementation of this aim-oriented empiricist view by the scientific community as the official, orthodox conception of science would correct the first blunder of the traditional Enlightenment.

But what of the *second* blunder? The task, here, is to generalize the methods of science appropriately so that they become progress-achieving methods that are, potentially, fruitfully applicable to *any*

---

[12] For a much more detailed exposition and defence of aim-oriented empiricism, and an account of the way aim-oriented empiricism solves long-standing problems in the philosophy of science such as problems of simplicity, induction and verisimilitude, see my (1998), chs. 1 and 3-6; and my (2007a), ch. 14.

worthwhile, problematic human endeavour. The task, in other words, is to generalize scientific rationality so that it becomes rationality *per se*, helping us to achieve what is of value whatever we may be doing.

Needless to say, scientists and philosophers, having failed to specify the methods of science properly, have also failed to arrive at the proper generalization of these methods. The best attempt known to me is that made by Karl Popper. According to Popper, science makes progress because it puts into practice the method of proposing theories as conjectures, which are then subjected to sustained attempted empirical refutation.[13] Popper argues that this can be generalized to form a conception of rationality – *critical rationalism* – according to which one seeks to solve problems quite generally by putting forward conjectures as to how a given problem is to be solved, these conjectures then being subjected to sustained *criticism* (criticism being a generalization of attempted empirical refutation in science).[14]

Popper's ideas about scientific method and how it is to be generalized are very important, and are a striking improvement over 18th century notions. In fact the "rational problem-solving" version of wisdom-inquiry sketched in chapter two depends quite essentially on Popper's critical rationalism. The first two rules of rational problem-solving, formulated in chapter two, are precisely the rules Popper arrives at as a result of generalizing his conception of scientific method.[15] Any problem-solving endeav-

---

[13] See Popper (1963), and (1959).

[14] "inter-subjective *testing* is merely a very important aspect of the more general idea of inter-subjective *criticism*, or in other words, of the idea of mutual rational control by critical discussion", Popper (1959), op. cit., p. 44, n *1. See also Popper (1963), pp. 193-200; (1976), pp. 115-6; and (1972), p. 119 and p. 243.

[15] Popper was too vehemently opposed to specialization to put forward rule (3). He failed to appreciate that the evils of rampant specialization can be corrected by implementing rule (4). Neither is included in Popper's problem-solving conception of critical rationalism. A typical anti-specialist comment of Popper is the following: "If the many, the specialists, gain the day, it will be the end of science as we know it – of great science. It will be a spiritual catastrophe comparable in its

our which persistently violates these two rules cannot hope to be rational. These two rules are, in other words, necessary conditions for rationality. (Academia as it is at present constituted, shaped by knowledge-inquiry, is damagingly irrational because it violates these two rules.)

Despite these virtues, Popper's ideas about scientific method and how it is to be generalized are nevertheless seriously defective. Popper's conception of scientific method is defective because it is a version of standard empiricism which, as we have seen, is untenable. It fails to identify the problematic aim of science properly, and thus fails to specify the need for science to improve its aims and methods as it proceeds. Popper's notion of critical rationalism is defective in an analogous way. It does not make improving aims and methods, when aims are problematic, an essential aspect of rationality.

If, however, we take the aim-oriented empiricist conception of scientific method as our starting point, and generalize that, the outcome is quite different. It is not just in science that aims are problematic; this is the case in life too, either because different aims conflict, or because what we believe to be desirable and realizable lacks one or other of these features, or both. Above all, the aim of creating global civilization is inherently and profoundly problematic (a point to be elaborated in a moment). Quite generally, then, and not just in science, whenever we pursue a problematic aim we need to represent the aim as a hierarchy of aims, from the specific and problematic at the bottom of the hierarchy, to the general and unproblematic at the top. In this way we provide ourselves with a framework within which we may improve more or less specific and problematic aims and methods as we proceed, learning from our successes and failures in practice what aims are both of most value and realizable. This conception of rationality, arrived at by generalizing the progress-achieving methods of science of aim-oriented empiricism, may be called *aim-oriented rationality*. Aim-oriented rationality improves on Popper's critical rationality in stressing the vital need, when one's

---

consequences to nuclear armament" Popper (1994), p. 72. For more expressions of Popper's opposition to specialization, see my (2010c).

aim is problematic, of representing the aim in the form of a hierarchy, thereby facilitating the *improvement* of specific, problematic aims and methods (low down in the hierarchy) as one proceeds.

The first step, in putting aim-oriented rationality into practice, is to try to make explicit both the actual problematic aim of the enterprise in question, and what the aim ought to be. This first step may be extraordinarily difficult to achieve. It has not yet been achieved as far as science is concerned. The official view is still that the actual and ideal intellectual aim of natural science is *knowledge of truth* (nothing being presupposed about the truth). The decisive refutation of this standard empiricist view, which I first published long ago in 1974 and developed in many publications since then,[16] and which I spelled out in the last chapter has, so far, made not the slightest dent in the scientific community's views about the aims and methods of science. Standard empiricism is still taken for granted.

Once this first step has been taken, the next step is to try to *improve* the real, problematic aim by solving the problems associated with this aim. This might involve reducing the gap between what we take to be the real and ideal aims, by modifying one or other, or both. (Sometimes what we are actually doing may be better, in some respects at least, than what we think we ought to be doing.) Our actual aim may need to be improved because, as it stands, it is unrealizable, or not as desirable, as valuable to achieve, as we suppose, or both. An aim may be less desirable than we suppose because it conflicts with other desirable aims in ways we did not anticipate, or has all sorts of undesirable, unforeseen consequences. We need to ask why the endeavour in question has the aim that it does have, both in the historical sense of how this aim came to be adopted, and in the rationalistic sense of for what further aim this aim is being pursued. Answers to these questions may reveal defects in the aim that is being pursued.

Finally, if aims of the endeavour in question seem to be permanently problematic, we need to represent the aim, and associated methods, in the form of a nested hierarchy of aims and

---

[16] See note 16 of the last chapter.

methods, aims becoming less and less specific and substantial as we go up the hierarchy, and thus less and less problematic and in need of revision. In this way, as I have already mentioned, we may create a framework of relatively unproblematic aims and methods within which much more specific and problematic aims, and associated methods, can be improved as we proceed, as we act.

Let me now illustrate these general points by considering the difficulties, and the potential benefits, of putting aim-oriented rationality into practice in the specific case of science.

In the case of science, a number of aims need to be considered.

There is the aim of theoretical physics of discovering the underlying unity inherent in all physical phenomena – this unity being presupposed to exist. We might call this, not the search for *truth* as such, but rather the search for *explanatory* truth, the truth being presupposed to be explanatory, or physically comprehensible, in some way or other. Because of the highly problematic character of this aim, it needs to be represented in the form of a hierarchy of increasingly unproblematic aims in the manner indicated in Diagram 4 of chapter 5.

The aim of discovering *explanatory truth* is, however, a special case of the more general aim of discovering, of improving our knowledge of, *valuable truth*. This is, if anything, even more problematic. Of value to whom? In what way? When? Who decides? Three permanent problems beset this aim of improving knowledge of valuable truth. First, two very different kinds of value are involved in science: cultural or intellectual on the one hand, and practical or technological on the other. Both are vital. But it is extraordinarily difficult to decide between the value of enhancing our understanding of the nature of the universe on the one hand and, let us say, ameliorating suffering and saving lives on the other. If these two kinds of value are not made explicit (because of the influence of standard empiricism) there is always the danger that research is conducted which leads to new knowledge but which realizes neither value even though much knowledge of truth is acquired Second, there is always the danger that scientists, in deciding what scientific aims to pursue, are governed by considerations internal to science itself (which can include factors such as scientific kudos and fashion), and ignore considerations of human need and suffering – above all the needs of the poor of the world. Those whose plight is the greatest are often also those least able to voice their plight. Third, it is important to appreciate that modern scientific and technological research can be extraordinarily expensive. He who pays the piper tends to call the tune. Inevitably, there will be the tendency for the priorities of research to come to reflect the interests of the wealthy and powerful rather than the needs of the poor.

But the aim of acquiring knowledge of valuable truth is not an end in itself. Even though this may not always be appreciated, it is pursued in order that knowledge and understanding of truth of value may be *used* by people – culturally or practically – to enhance the quality of life. Science, in the end, at its best, has a humanitarian goal: to contribute towards the enhancement of the quality of human life (and this is true of both aspects of science, the cultural and the practical). But this humanitarian or social aim is even more profoundly problematic. And all sorts of doubts may be raised about the success of science in achieving this aim.

Much more needs to be said about the diverse and problematic aims of science. I have said enough, I hope, to indicate just how important it is to throw the aims of science open to sustained imaginative and critical exploration in an attempt to *improve* the aims of science – how important it is to apply aim-oriented rationality to science, and how inadequate standard empiricism is in its characterization of the aims of science. Putting aim-oriented rationality into scientific practice is doubly important: it is needed to correct serious defects – intellectual and humanitarian – in science itself. And it is needed so that science might be what it should be, a working example of what it is to put aim-oriented rationality into practice, and of why it is so important to do. Aim-oriented empiricist science could be taken as a model, a paradigm, for aim-oriented rationalistic action quite generally.

So much for the second blunder, and how it is to be put right. We come now to the *third* blunder.

This is by far the most serious of the three blunders made by the traditional Enlightenment. The basic Enlightenment idea, after all, is to learn from our solution to the first great problem of learning how to solve the second problem – to learn, that is, from scientific progress how to make social progress towards an enlightened world. Putting this idea into practice involves getting appropriately generalized progress-achieving methods of science *into social life itself!* It involves getting progress-achieving methods into our institutions and ways of life, into government, industry, agriculture, commerce, international relations, the media, the arts, education. But in sharp contrast to all this, the traditional Enlightenment sought to apply generalized scientific method, not to social *life*, but merely to social *science!* Instead of helping humanity learn how to become more *civilized* by rational means, the traditional Enlightenment sought merely to help social scientists improve *knowledge* of social phenomena. The outcome is that today academic inquiry devotes itself to acquiring knowledge of natural and social phenomena, but does not attempt to help humanity learn how to become more civilized. This is the blunder that is at the root of our current failure to have solved the second great problem of learning.[17]

In order to correct the third blunder of the traditional Enlightenment, then, social inquiry and the humanities need to take up the *methodological, social,* and even *political* task of helping to get into *social life itself,* into all our other institutions besides science, the progress-achieving methods of aim-oriented rationalism arrived at by generalizing the aim-oriented empiricist methods of natural science. Social inquiry is not primarily social *science.* It is not primarily concerned to acquire knowledge of social phenomena – although this is indeed an important matter of secondary concern. The intellectually fundamental aim of social inquiry is to help humanity build into institutions and social life quite generally the progress-achieving methods of aim-oriented rationality (arrived at by generalizing the progress-achieving methods of science as indicated above). Social inquiry (sociology, economics, anthropology and the rest) is thus social *methodology* or social *philosophy.* Its task is to help diverse valuable human endeavours and institutions gradually improve aims and methods

---

[17] For a discussion of the extent to which the traditional Enlightenment dominates academic inquiry see my (2007a), chapters 6 and 12.

so that they may make more worthwhile contributions to human life.

Above all, social inquiry and the humanities need to help us learn how to put aim-oriented rationality into practice in seeking to make progress towards a genuinely civilized world. As I have already remarked, the aim of achieving global civilization is inherently and profoundly problematic. People have very different ideas as to what does constitute civilization. Most views about what constitutes an ideally civilized society have been unrealizable *and* profoundly undesirable. People's interests, values and ideals clash. Even values that, one may hold, ought to be a part of civilization may clash. Thus freedom and equality, even though inter-related, may nevertheless clash. It would be an odd notion of individual freedom which held that freedom is for some and not for others; and yet if equality is pursued too single-mindedly this will undermine individual freedom, and will even undermine equality, in that a privileged class will be required to enforce equality on the rest, as in the old Soviet Union. A basic aim of legislation for civilization, we may well hold, ought to be increase freedom by restricting it: this brings out the inherently paradoxical character of the aim of achieving civilization.

The inherently problematic character of the aim of achieving world civilization makes it imperative that here, above all, we proceed in accordance with the edicts of aim-oriented rationality. This means at least that we represent the aim at a number of levels, from the specific and highly problematic to the unspecific and unproblematic. Thus, at a fairly specific level, we might specify civilization to be a state of affairs in which there is an end to war, dictatorships, population growth, extreme inequalities of wealth, and the establishment of democratic, liberal world government and a sustainable world industry and agriculture. At a rather more general level we might specify civilization to be a state of affairs in which everyone shares equally in enjoying, sustaining and creating what is of value in life *in so far as this is possible*. At a still more general level, civilization might be specified simply as that realizable world order we should seek to attain in the long term: see diagram 5.

As a result of building into our institutions and social life such a hierarchical structure of aims and associated methods, it becomes possible for us to develop and assess rival philosophies of life as a part of social life, somewhat as theories are developed and assessed within science. Such a hierarchical methodology provides a framework within which competing views about what our aims and methods in life should be – competing religious, political and moral views – may be cooperatively assessed and tested against broadly agreed, unspecific aims (high up in the

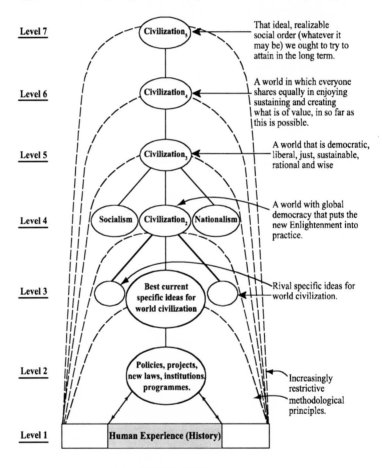

Diagram 5
Aim-Oriented Rationality Applied to the Task of Making
Progress towards a Civilized World

180

hierarchy of aims) and the experience of personal and social life. There is the possibility of cooperatively and progressively improving *such philosophies of life* (views about what is of value in life and how it is to be achieved) much as *theories* are cooperatively and progressively improved in science. In science, ideally, theories are critically assessed with respect to each other, with respect to metaphysical ideas concerning the comprehensibility of the universe, and with respect to *experience* (observational and experimental results). In a somewhat analogous way, diverse philosophies of life may be critically assessed with respect to each other, with respect to relatively uncontroversial, agreed ideas about aims and what is of value, and with respect to *experience* – what we do, achieve, fail to achieve, enjoy and suffer – the aim being to improve philosophies of life (and more specific philosophies of more specific enterprises within life such as government, education or art) so that they offer greater help with the realization of what is of value in life. This hierarchical methodology is especially relevant to the task of resolving conflicts about aims and ideals, as it helps disentangle agreement (high up in the hierarchy) and disagreement (more likely to be low down in the hierarchy). The hope in all this is that, as a consequence of the two-way interaction between our lives and our philosophies of life (our policies, our plans, our political programmes), as our philosophies of life improve, so our lives improve as well.

The upshot of the whole argument – the argument of chapter two, appealing to problem-solving rationality, and the argument just indicated, appealing to aim-oriented rationality and correcting the three mistakes of the 18th century Enlightenment – is that, not just social inquiry, but the whole academic enterprise needs to change. The primary task of academia as a whole ought to be to help humanity solve its problems of living in increasingly rational, cooperative, enlightened ways, thus helping humanity become more civilized. This task would be intellectually more fundamental than the scientific task of acquiring knowledge. Social inquiry would be intellectually more fundamental than physics. As I have already remarked, academia would be a kind of people's civil service, doing openly for the public what actual civil services are supposed to do in secret for governments. Academia would have just sufficient power (but no more) to retain its independence from government, industry, the press, public opinion, and other centres of power and influence in the social world. It would seek to learn from, educate, and argue with the great social world beyond, but would not dictate. Academic thought would be pursued as a specialized, subordinate part of what is really important and fundamental: the thinking that goes on, individually, socially and institutionally, in the social world, guiding individual, social and institutional actions and life. The fundamental intellectual and humanitarian aim of inquiry would be to help humanity acquire wisdom – wisdom being the capacity to realize (apprehend and create) what is of value in life, for oneself and others, wisdom thus including knowledge and technological know-how but much else besides. Scientific and technological research would be conducted in such a way that knowledge acquired contributes to wisdom.

Wisdom-inquiry, because of its greater rigour,[18] has intellectual standards that are, in important respects, different from those of knowledge-inquiry. Whereas knowledge-inquiry demands that emotions and desires, values, human ideals and aspirations, philosophies of life be excluded from the intellectual domain of inquiry, wisdom-inquiry requires that they be included. In order to discover what is of value in life it is essential that we attend to our feelings and desires. But not everything we desire is desirable, and not everything that feels good is good. Feelings, desires and values need to be subjected to critical scrutiny. And of course feelings, desires and values must not be permitted to influence judgements of factual truth and falsity. Wisdom-inquiry embodies a synthesis of traditional rationalism and romanticism. It includes elements from both, and it improves on both. It incorporates romantic ideals of integrity, having to do with motivational and emotional honesty, honesty about desires and aims; and at the same time it incorporates traditional rationalist ideals of integrity, having to do with respect for objective fact, knowledge, and valid

---

[18] Wisdom-inquiry puts the four rules of problem-solving rationality, and the edicts of aim-pursuing rationality into practice, whereas knowledge-inquiry, as we have seen, does not. It is this that renders wisdom-inquiry more rigorous than knowledge-inquiry.

argument. Traditional rationalism takes its inspiration from science and method; romanticism takes its inspiration from art, from imagination, and from passion. Wisdom-inquiry holds art to have a fundamental rational role in inquiry, in revealing what is of value, and unmasking false values; but science, too, is of fundamental importance. What we need, for wisdom, is an interplay of sceptical rationality and emotion, an interplay of mind and heart, so that we may develop mindful hearts and heartfelt minds. It is time we healed the great rift in our culture, so graphically depicted by C. P. Snow.[19]

Aim-oriented rationality may seem almost banal in its obvious and elementary character. This might mislead one into thinking it is an easy matter to put it into practice in science, in academia, in education, in politics, industry, commerce, international relations, and into life quite generally. Nothing could be further from the truth. Fear of the unfamiliar, rigid habits of thought and of living, prejudice and, above all vested interests, will make it very difficult indeed to implement aim-oriented rationality. This is the case even in a field such as theoretical physics, where it might seem a relatively easy matter to put it explicitly into practice (especially as it is already put into practice covertly). My thirty years of trying to get the idea accepted by the physics community have failed to get it noticed even by philosophers of physics, let alone physicists themselves. There is fierce resistance to acknowledging that science as a whole has to accept, as articles of faith – even if *rational* faith – untestable, metaphysical doctrines about the nature of the universe, before the empirical method of science can get underway. As I have already indicated, it is important, for rather general reasons, that theoretical physics *does* put aim-oriented empiricism into practice. If it did, it would constitute an exemplary case of aim-oriented rationality being implemented from which attempts to put aim-oriented rationality into practice elsewhere, in more challenging contexts, could learn. For these other contexts – politics, industry, the law, the media, international relations – pose problems far more intractable than the particular case of physics, or natural science more generally. Here, hypocrisy, misrepresentation of aims and ideals, is endemic, a

---

[19] Snow (1964).

standard strategy for holding onto power, wealth and status. Fierce opposition to even the most modest proposals for some sustained scrutiny of aims must be expected.

Cultural Implications of Wisdom-Inquiry

It might seem, from what has been said so far, that wisdom-inquiry may do better justice than knowledge-inquiry to the practical aspects of inquiry – its capacity to help people achieve desirable ends in life – but only at the expense of neglecting the "pure" or "intellectual" aspects of inquiry, the value of science and scholarship when pursued for their own sake, in other words.

The actual situation is all the other way round. Wisdom-inquiry does better justice to *both* aspects of inquiry, pure and applied.

To begin with, aim-oriented empiricism (an integral part of wisdom-inquiry) does better justice to theoretical physics pursued for its own sake than does standard empiricism (an integral part of knowledge-inquiry). This is because aim-oriented empiricism emphasizes the search for *understanding* in a way in which standard empiricism does not. Physics pursued in accordance with aim-oriented empiricism becomes much more like natural philosophy – what physics was for Newton and his contemporaries, an intermingling of testable physics, metaphysics and methodology, even philosophy. It is a passionate quest to understand the nature of the universe. Physics pursued in accordance with standard empiricism is in danger of becoming merely the business of predicting more and more phenomena more and more accurately. This is in part because standard empiricism banishes metaphysics from physics – and thus banishes explicit discussion of problems of understanding. It is also because, whereas aim-oriented empiricism demands that acceptable theories be sufficiently empirically successful *and* unified, standard empiricism demands only sufficient empirical success. There are additional requirements of simplicity, it is true, but exactly what these requirements amount to is left obscure, and they are weak in that they cannot over-ride empirical requirements. Aim-oriented empiricism, by contrast, makes crystal clear what unified *means*, and insists that the demand for unity may well over-ride empirical considerations. In clarifying and strengthening the demand for theoretical unity in this way, aim-oriented empiricism in effect insists that acceptable theories must *explain* and embody *understanding*. It is not enough merely to *predict*.

These differences are strikingly illustrated by the case of orthodox quantum theory. As we saw in the last chapter, this theory, notoriously, fails to explain, and enable us to understand the quantum domain because it fails to solve the quantum wave/particle problem – and thus fails to say clearly what sort of entities quantum objects, such as electrons, photons and atoms, really are. Because of this failure, orthodox quantum theory is a theory about the results of performing measurements on quantum entities, not a theory about these entities as such. But this in turn means that orthodox quantum theory is severely ad hoc. It is made up of quantum postulates plus some part of classical physics to describe measurement. This is a severely disunified theory. Its failure to embody understanding is directly linked to its disunity, its severely ad hoc character, in turn linked to its failure to solve the crucial wave/particle problem.

Viewed from the perspective of standard empiricism, orthodox quantum theory is a supremely acceptable theory because of its astonishing empirical success. Its disunity, its failure to provide understanding, does not provide grounds for rejecting the theory. But viewed from the perspective of aim-oriented empiricism, the theory is unacceptably ad hoc and disunified. Its failure to embody understanding *does* provide grounds for rejection. Furthermore, the kind of metaphysical reasoning which aim-oriented empiricism encourages, and standard empiricism fails to encourage, readily shows how the wave-particle problem can be solved, and a more unified version of the theory can be developed, as I have shown elsewhere.[20] All this strikingly bears out the point that aim-oriented empiricism does far better justice than standard empiricism to the pursuit of understanding in physics, to the value of physics pursued for its own sake.

There are additional, more general ways in which wisdom-inquiry does better justice than knowledge-inquiry to inquiry

---

[20] See my (1998), ch. 7, and further references therein. See also my (2004b) and (2010b).

pursued for its own sake.

From the standpoint of the intellectual or cultural aspect of inquiry, what really matters is the desire that people have to see, to know, to understand, the passionate curiosity that individuals have about aspects of the world, and the knowledge and understanding that people acquire and share as a result of actively following up their curiosity. An important task for academic thought in universities is to encourage non-professional thought to flourish outside universities. As Einstein once remarked "Knowledge exists in two forms – lifeless, stored in books, and alive in the consciousness of men. The second form of existence is after all the essential one; the first, indispensable as it may be, occupies only an inferior position."[21]

Wisdom-inquiry is designed to promote all this in a number of ways. It does so as a result of holding thought, at its most fundamental, to be the personal thinking we engage in as we live. It does so by recognizing that acquiring knowledge and understanding involves articulating and solving personal problems that one encounters in seeking to know and understand. It does so by recognizing that passion, emotion and desire, have a rational role to play in inquiry, disinterested research being a myth. Again, as Einstein has put it "The most beautiful experience we can have is the mysterious. It is the fundamental emotion which stands at the cradle of true art and true science. Whoever does not know it and can no longer wonder, no longer marvel, is as good as dead, and his eyes are dimmed."[22]

Knowledge-inquiry, by contrast, all too often fails to nourish "the holy curiosity of inquiry",[23] and may even crush it out altogether. Knowledge-inquiry gives no rational role to emotion and desire; passionate curiosity, a sense of mystery, of wonder, have no place, officially, within the rational pursuit of knowledge. The intellectual domain becomes impersonal and split off from personal feelings and desires; it is difficult for "holy curiosity" to flourish in such circumstances. Knowledge-inquiry hardly

---

[21] Einstein (1973), p. 80.
[22] Einstein (1973), p. 11.
[23] Einstein (1949), p. 17.

encourages the view that inquiry at its most fundamental is the thinking that goes on as a part of life; on the contrary, it upholds the idea that fundamental research is highly esoteric, conducted by physicists in contexts remote from ordinary life. Even though the aim of inquiry may, officially, be *human* knowledge, the personal and social dimension of this is all too easily lost sight of, and progress in knowledge is conceived of in impersonal terms, stored lifelessly in books and journals. Rare is it for popular books on science to take seriously the task of exploring the fundamental problems of a science in as accessible, non-technical and intellectually responsible a way as possible. [24] Such work is not highly regarded by knowledge-inquiry, as it does not contribute to "expert knowledge". The failure of knowledge-inquiry to take seriously the highly problematic nature of the aims of inquiry leads to insensitivity as to what aims are being pursued, to a kind of institutional hypocrisy. Officially, knowledge is being sought "for its own sake", but actually the goal may be immortality, fame, the flourishing of one's career or research group, as the existence of bitter priority disputes in science indicates. Education suffers. Science students are taught a mass of established scientific knowledge, but may not be informed of the *problems* which gave rise to this knowledge, the problems which scientists grappled with in creating the knowledge. Even more rarely are students encouraged themselves to grapple with such problems. And rare, too, is it for students to be encouraged to articulate their own problems of understanding that must, inevitably arise in absorbing all this information, or to articulate their instinctive criticisms of the received body of knowledge. All this tends to reduce education to a kind of intellectual indoctrination, and serves to kill "holy curiosity". Officially, courses in universities divide up into those that are vocational, like engineering, medicine and law, and those that are purely educational, like physics, philosophy or history. What is not noticed, again through insensitivity to problematic aims, is that the supposedly purely educational are actually vocational as well: the student is being trained to be an academic physicist, philosopher or historian, even though only a minute

---

[24] A recent, remarkable exception is Penrose (2004).

187

percentage of the students will go on to become academics. Real education, which must be open-ended, and without any pre-determined goal, rarely exists in universities, and yet few notice.[25]

In order to enhance our understanding of persons as beings of value, potentially and actually, we need to understand them empathetically, by putting ourselves imaginatively into their shoes, and experiencing, in imagination, what they feel, think, desire, fear, plan, see, love and hate. For wisdom-inquiry, this kind of empathic understanding is rational and intellectually fundamental. Articulating problems of living, and proposing and assessing possible solutions is, we have seen, the fundamental intellectual activity of wisdom-inquiry. But it is just this that we need to do to acquire empathic understanding. Social inquiry, in tackling problems of living, is also promoting empathic understanding of people. Empathic understanding is essential to wisdom. Elsewhere I have argued, indeed, that empathic understanding plays an essential role in the evolution of consciousness. It is required for cooperative action, and even for science.[26]

Granted knowledge-inquiry, on the other hand, empathic understanding hardly satisfies basic requirements for being an intellectually legitimate kind of explanation and understanding.[27] It has the status merely of "folk psychology", on a par with "folk physics".

The intellectually fundamental character of empathic or "personalistic" understanding will turn out to be of great importance when we come to the question of free will, to be discussed in the next chapter.

Objections

I now consider, briefly, some objections that may be raised against my claim that wisdom-inquiry is both more rigorous and more humanly desirable than knowledge-inquiry.

---

[25] These considerations are developed further in my (1976a), (1984) and (2004a).

[26] For fuller expositions of such an account of empathic understanding see my (1984), pp. 171-189 and ch. 10; and (2001), chs. 5-7 and 9.

[27] See my (1984), pp. 183-185.

It may be objected that the traditional Enlightenment does not dominate current academic inquiry to the extent that I have assumed. But grounds for holding that it does are given in chapter six of my *From Knowledge to Wisdom*. There I looked at the following: (1) books about the modern university; (2) the philosophy and sociology of science; (3) statements of leading scientists; (4) Physics Abstracts; (5) Chemistry, Biology, Geo and Psychology Abstracts; (6) journal titles and contents; (7) books on economics, sociology and psychology; (8) philosophy. In 1984, the year *From Knowledge to Wisdom* was published, there can be no doubt whatsoever that the traditional Enlightenment (or "the philosophy of knowledge" as I called it in the book) dominated academic inquiry.

Have things changed since then? The revolution advocated by *From Knowledge to Wisdom*, and argued for here, has not occurred. There is still, amongst the vast majority of academics today, no awareness at all that a more intellectually rigorous and humanly valuable kind of inquiry than that which we have at present, exists as an option. In particular, social inquiry continues to be taught and pursued as social *science*, and not as social *methodology*. Recently I undertook an examination, at random, of thirty-four introductory books on sociology, published between 1985 and 1997. Sociology, typically, is defined as "the scientific study of human society and social interactions",[28] as "the *systematic, sceptical study of human society*",[29] or as having as its basic aim "to understand human societies and the forces that have made them what they are".[30] Some books take issue with the idea that sociology is the *scientific* study of society, or protest at the male dominated nature of sociology.[31] Nowhere did I find a hint of the idea that a primary task of sociology, or of social inquiry more generally, might be to help build into the fabric of social life progress-achieving methods, generalized from those of science,

---

[28] Tischler (1996), p. 4.
[29] Macionis and Plummer (1997), p. 4.
[30] Lenski et al. (1995), p. 5.
[31] See, for example, Abott and Wallace (1990), p. 3 and p. 1.

designed to help humanity resolve its conflicts and problems of living in more cooperatively rational ways than at present.[32]

The tackling of problems of living rather than problems of knowledge does of course go on within the academic enterprise as it is at present constituted, within such disciplines as economics, development studies, policy studies, peace studies, medicine, agriculture, engineering, and elsewhere. In chapter nine I will indicate some recent developments in academia which can perhaps be interpreted as amounting to the first faltering steps away from knowledge-inquiry towards wisdom-inquiry. None of this, however, tells against the point that the primary task of academic inquiry at present is still, first, to acquire knowledge and technological know-how, and then, second, to apply it to help solve problems of living. It does not, in other words, tell against the point that it is knowledge-inquiry and the traditional Enlightenment that are still, at the time of writing, the dominant influence on the nature, the aims and methods, the whole character and structure of academic inquiry.

It may be objected that it is all to the good that the academic enterprise today does give priority to the pursuit of knowledge over the task of promoting wisdom and civilization. Before problems of living can be tackled rationally, knowledge must first be acquired.[33]

I have six replies to this objection.

First, even if the objection were valid, it would still be vital for a kind of inquiry designed to help us build a better world to include rational exploration of problems of living, and to ensure that this guides priorities of scientific research (and is guided by the results of such research).

Second, the validity of the objection becomes dubious when we take into account the considerable success people met with in solving problems of living in a state of extreme ignorance, before

---

[32] See also my (2007a), ch. 6.

[33] This is the objection that most academics will wish to raise against my "from knowledge to wisdom" thesis. It will be made by all those who hold that academic inquiry quite properly seeks to make a contribution to human welfare by, first, acquiring knowledge and then, secondarily, applying it to help solve human problems.

the advent of science. We still today often arrive at solutions to problems of living in ignorance of relevant facts.

Third, the objection is not valid. In order to articulate problems of living and explore imaginatively and critically possible solutions (in accordance with Popper's conception of rationality) we need to be able to act in the world, imagine possible actions and share our imaginings with others: in so far as some common sense knowledge is implicit in all this, such knowledge is required to tackle rationally and successfully problems of living. But this does not mean that we must give intellectual priority to acquiring new relevant knowledge before we can be in a position to tackle rationally our problems of living.

Fourth, simply in order to have some idea of what kind of knowledge or know-how it is *relevant* for us to try to acquire, we must *first* have some provisional ideas as to what our problem of living is and what we might do to solve it. Articulating our problem of living and proposing and critically assessing possible solutions needs to be intellectually prior to acquiring relevant knowledge simply for this reason: we cannot know what new knowledge it is *relevant* for us to acquire until we have at least a preliminary idea as to what our problem of living is, and what we propose to do about it. A slight change in the way we construe our problem may lead to a drastic change in the kind of knowledge it is relevant to acquire: changing the way we construe problems of health, to include *prevention* of disease (and not just curing of disease) leads to a dramatic change in the kind of knowledge we need to acquire (importance of exercise, diet etc.). Including the importance of avoiding *pollution* in the problem of creating wealth by means of industrial development leads to the need to develop entirely new kinds of knowledge.

Fifth, relevant knowledge is often hard to acquire; it would be a disaster if we suspended life until it had been acquired. Knowledge of how our brains work is presumably highly relevant to all that we do but clearly, suspending rational tackling of problems of living until this relevant knowledge has been acquired would not be a sensible step to take. It would, in any case, make it impossible for us to acquire the relevant knowledge (since this requires scientists to act in doing research). Scientific research is itself a kind of action carried on in a state of relative ignorance.

Sixth, the capacity to act, to live, more or less successfully in the world, is more fundamental than (propositional) knowledge. Put in Rylean terms, 'knowing how' is more fundamental than 'knowing that'.[34] All our knowledge is but a development of our capacity to act. Dissociated from life, from action, knowledge stored in libraries is just paper and ink, devoid of meaning. In this sense, problems of living are more fundamental than problems of knowledge (which are but an aspect of problems of living); giving intellectual priority to problems of living quite properly reflects this point.[35]

As I have already stressed, a kind of inquiry that gives priority to tackling problems of knowledge over problems of living violates the most elementary requirements of rationality conceivable. If the basic task is to help humanity create a better world, then the problems that need to be solved are, primarily, problems of living, problems of action, not problems of knowledge. This means that to comply, merely, with Popper's conception of critical rationalism (or problem-solving rationality) discussed above, the basic intellectual tasks need to be (1) to articulate problems of living, and (2) to propose and critically assess possible solutions, possible more or less cooperative human *actions*. (1) and (2) are excluded, or marginalized, by a kind of inquiry that gives priority to the task of solving problems of knowledge. And the result will be a kind of inquiry that fails to create a reservoir of imaginative and critically examined ideas for the resolution of problems of living, and instead develops knowledge often unrelated to, or even harmful to, our most basic human needs.

It may be objected that in employing aim-oriented rationality in an attempt to help create a more civilized world, in the way indicated above, the new Enlightenment falls foul of Popper's strictures against what he calls Utopian social engineering. This, for Popper, "aims at remodelling the 'whole of society' in accordance with a definite plan or blueprint; it aims at 'seizing the

---

[34] Ryle (1949), ch. II.
[35] For a development of this point, see my (1984), pp. 174-181.

key positions' and at 'extending the power of the State . . . until the State becomes nearly identical with society', and it aims, furthermore, at controlling from these 'key positions' the historical forces that mould the future of the developing society".[36] Popper contrasts Utopian engineering, which he rejects as irrational and disastrous, with piecemeal social engineering, which he advocates as rational and humanitarian. This is characterized as follows:

> Even though he (the piecemeal engineer) may perhaps cherish some ideals which concern society 'as a whole' – its general welfare perhaps – he does not believe in the method of re-designing it as a whole. Whatever his ends, he tries to achieve them by small adjustments and re-adjustments which can be continually improved upon. His ends may be of diverse kinds, for example, the accumulation of wealth or of power by certain individuals, or by certain groups; or the distribution of wealth and power; or the protection of certain 'rights' of individuals or groups, etc. Thus public or political (piecemeal) engineering may have the most diverse tendencies, totalitarian as well as liberal . . . The piecemeal engineer knows, like Socrates, how little he knows. He knows that we can learn from our mistakes. Accordingly, he will make his way, step by step, carefully comparing the results expected with the results achieved, and always on the look-out for the unavoidable unwanted consequences of any reform; and he will avoid undertaking reforms of a complexity and scope which make it impossible for him to disentangle causes and effects, and to know what he is really doing.[37]

The New Enlightenment project of endeavouring to make progress towards a good world by putting aim-oriented rationality into practice in personal, social, institutional and global life may

---

[36] Popper (1962), p. 67.
[37] Ibid, pp. 66-7.

seem to be somewhat similar to Popper's Utopian social engineering, and hence to fall foul of Popper's criticisms.

I have three replies to this objection. First, to the extent that piecemeal social engineering, of the kind advocated by Popper, is indeed the rational way to make progress towards a more civilized world, this will be advocated by the New Enlightenment. Second, when we take into account the unprecedented *global* nature of many of our most serious problems, indicated above, (the outcome of solving the first great problem of learning but failing to solve the second), we may well doubt that piecemeal social engineering is sufficient. Third, Popper's distinction between piecemeal and Utopian social engineering is altogether too crude: it overlooks entirely what has been advocated here, aim-oriented rationalistic social engineering, with its emphasis on developing increasingly cooperatively rational resolutions of human conflicts and problems in full recognition of the inherently problematic nature of the aim of achieving greater civilization.[38]

All those to any degree influenced by Romanticism and the counter-Enlightenment will object strongly to the idea that we should learn from scientific progress how to achieve social progress towards civilization; they will object strongly to the idea of allowing conceptions of rationality, stemming from science, to dominate in this way, and will object even more strongly to the idea, inherent in the new Enlightenment, that we need to create a more aim-oriented rationalistic social world.[39]

Directed at the traditional Enlightenment, objections of this kind have some validity; but directed at the new Enlightenment, they have none. As I have emphasized, aim-oriented rationality amounts to a synthesis of traditional rationalist and romantic ideals, and not to the triumph of the first over the second. In giving priority to the realization of what is of value in life, and in emphasizing that rationality demands that we seek to improve aims

---

[38] For further discussion see my (1984) pp. 189-198; (2007a), pp. 213-221 and 338-344.

[39] For literature protesting against the influence of scientific rationality in various contexts and ways, see for example: Berlin (1999); Laing (1965); Marcuse (1964): Roszak (1973); Berman (1981); Schwartz (1987); Feyerabend (1978) and (1987); Appleyard (1992).

as we proceed, the new Enlightenment requires that rationality integrates traditional Rationalist and Romantic values and ideals of integrity. Imagination, emotion, desire, art, empathic understanding of people and culture, the imaginative exploration of aims and ideals, which tend to be repudiated as irrational by traditional Rationalism, but which are prized by Romanticism, are all essential ingredients of aim-oriented rationality. Far from crushing freedom, spontaneity, creativity and diversity, aim-oriented rationality is essential for the desirable flourishing of these things in life.[40]

Many historians and sociologists of science deny that there is any such thing as scientific method or scientific progress, and will thus find the basic idea of this chapter absurd.[41] These writers are encouraged in their views by the long-standing failure of scientists and philosophers of science to explain clearly what scientific method is, and how it is to be justified. This excuse for not taking scientific method and progress seriously is, however, no longer viable: as I have indicated above, reject standard empiricism, and it becomes clear how scientific method and progress are to be characterized and justified, in a way which emphasizes the rational interplay between evolving knowledge and evolving aims and methods of science.[42] In a world dominated by the products of scientific progress it is quixotic in the extreme to deny that such progress has taken place.

Finally, those of a more rationalist persuasion may object that science is too different from political life for there to be anything worthwhile to be learnt from scientific success about how to

---

[40] See my (1984) pp. 63-4, pp. 85-91 and pp. 117-118, for further discussion of this issue. See also my (1976a), especially chs. 1 and 8-10.
[41] Bloor (1976); B. Barnes and D. Bloor (1981); Latour (1987); Feyerabend (1978) and (1987). These authors might protest that they do not deny scientific knowledge, method, progress or rationality as such, but deny, merely, that the sociology of knowledge can legitimately appeal to such things, or deny extravagant claims made on behalf of these things. See, however, the sparkling criticism by Sokal and Bricmont (1998), ch. 4.
[42] See my (1998), especially chs 1-6; (2004a), especially the appendix; and my (2007a), ch. 14.

achieve social progress towards civilization.[43] (a) In science there is a decisive procedure for eliminating ideas, namely, empirical refutation: nothing comparable obtains, or can obtain, in the political domain. (b) In science experiments or trials may be carried out relatively painlessly (except, perhaps, when new drugs are being given in live trials); in life, social experiments, in that they involve people, may cause much pain if they go wrong, and may be difficult to stop once started. (c) Scientific progress requires a number of highly intelligent and motivated people to pursue science on the behalf of the rest of us, funded by government and industry; social progress requires almost *everyone* to take part, including the stupid, the criminal, the mad or otherwise handicapped, the ill, the highly unmotivated; and in general there is no payment. (d) Scientists, at a certain level, have an agreed, common objective: to improve knowledge. In life, people often have quite different or conflicting goals, and there is no general agreement as to what civilization ought to mean, or even whether it is desirable to pursue civilization in *any* sense. (e) Science is about fact, politics about value, the quality of life. This difference ensures that science has nothing to teach political action (for civilization). (f) Science is male-dominated, fiercely competitive, and at times terrifyingly impersonal;[44] this means it is quite unfit to provide any kind of guide for life.

Here, briefly, are my replies. (a) Some proposals for action can be shown to be unacceptable quite decisively as a result of experience acquired through attempting to put the proposal into action. Where this is not possible, it may still be possible to assess the merits of the proposal to some extent by means of experience. If assessing proposals for action by means of experience is much more indecisive than assessing scientific theories by means of experiment, then we need, all the more, to devote our care and attention to the former case. (b) Precisely because experimentation in life is so much more difficult than in science, it is vital that in life we endeavour to learn as much as possible from (i) experiments that we perform in our imagination, and (ii)

---

[43] N. Rescher (personal communication); Durant (1997).
[44] Harding (1986).

196

experiments that occur as a result of what actually happens. (c) Because humanity does not have the aptitude or desire for wisdom that scientists have for knowledge, it is unreasonable to suppose that progress towards global wisdom could be as explosively rapid as progress in science. Nevertheless progress in wisdom might go better than it does at present. (d) Cooperative rationality is only feasible when there is the common desire of those involved to resolve conflicts in a cooperatively rational way. (e) Aim-oriented rationality can help us improve our decisions about what is desirable or of value, even if it cannot reach decisions for us. (f) In taking science as a guide for life, it is the progress-achieving methodology of science to which we need to attend. It is this that we need to generalize in such a way that it becomes fruitfully applicable, potentially, to all that we do. That modern science is male-dominated, fiercely competitive, and at times terrifyingly impersonal should not deter us from seeing what can be learned from the progress-achieving methods of science - unless, perhaps, it should turn out that being male-dominated, fiercely competitive and impersonal is essential to scientific method and progress. (But this, I submit, is not the case.)

Conclusion

Having solved the first great problem of learning, it has become a matter of extreme urgency, as far as the future of humanity is concerned, that we discover how to solve the second problem. In order to do this we need to correct the three blunders of the traditional Enlightenment. This involves changing the nature of social inquiry, so that social *science* becomes social *methodology* or social *philosophy*, concerned to help us build into social life the progress-achieving methods of aim-oriented rationality, arrived at by generalizing the progress-achieving methods of science. It also involves, more generally, bringing about a revolution in the nature of academic inquiry as a whole, so that it takes up its proper task of helping humanity learn how to become wiser by increasingly cooperatively rational means. The scientific task of improving knowledge and understanding of nature becomes a part of the broader task of improving global wisdom.

197

If ever wisdom-inquiry, as sketched here (and as spelled out in more detail elsewhere[45]) is put into practice in schools, universities and research institutions throughout the world, humanity would at last have what it so urgently needs: a kind of inquiry rationally designed and devoted to helping the God-of-Cosmic-Value to flourish within the God-of-Cosmic-Power, in so far as it is able to do so.

---

[45] See my (1976a); (1984); (2004a); and (2007a). For critical assessments of wisdom-inquiry see McHenry (2009).

CHAPTER SEVEN

HOW CAN WE EXIST INSIDE THE GOD-OF-COSMIC-POWER?

Physics is about only a highly selected aspect of things - the causally efficacious aspect, that aspect which determines (perhaps probabilistically) the way events unfold. Physics makes no mention of the experiential – how things look, feel, smell, or sound, or how it is to *be* a bit of the physical universe (this body, this brain) – first because none of this is needed to predict and explain any physical phenomena, and secondly because, if physical theory is extended to include all this experiential stuff, the theory would become so horrendously complex it would cease to be explanatory. Omitting all reference to the experiential is the price that must be paid to have the marvellously explanatory theories of physics that we do have – Newtonian theory, James Clerk Maxwell's theory of the electromagnetic field, Einstein's theory of general relativity, quantum theory, quantum electrodynamics and the so-called "standard model", the current best theory we have of fundamental particles and the forces between them. The God-of-Cosmic-Power is just that aspect, that slice, of all that which exists, which we must pick out if we are to explain and understand why events occur as they do. Physics is about the explanatory skeleton of the world with all the experiential flesh omitted.

This is what I argued for in chapter three. If correct, it goes some way towards enabling us to understand how the experiential God-of-Cosmic-Value can coexist with the physical God-of-Cosmic-Power. The silence of physics about the experiential provides no grounds whatsoever for holding that the experiential does not really exist.

But much more than this is required to reconcile the God-of-Cosmic-Value with the God-of-Cosmic-Power. There is still the grave problem of showing how it is possible that *we* exist in the physical universe. We persons are not merely passively experiencing beings - vessels for passing sensations of sights and sounds and smells. We *do* things in this world, we *act*, we initiate actions and are responsible for what we do. It is of the very essence of what we are that we act. Even our thinking, our imagining, our inner world of consciousness, our very identity as a conscious being, is a kind of action, the product of inner imaginary action. Bereft of the capacity to *do*, to *act*, we are nothing, we cease to exist altogether.

But if the physical God-of-Cosmic-Power exists, everything that occurs - or at least the physical aspect of everything that occurs - is capable (in principle) of being explained and understood in purely physical terms. Everything that we do, say, think, feel, all our deeds throughout our life from the most momentous to the smallest flicker of an eyelid are the inexorable outcome of prior physical states of the universe. Our entire life is just the outcome of fundamental physical entities – electrons, quarks, superstrings, or whatever – interacting with one another in accordance with precise physical law. We may feel we exist and act; our life certainly seems charged with human significance, experience, emotion, struggle, plans, intentions, deeds that sometimes succeed and sometimes fail. But all this is, it seems, a hollow charade. Behind the scenes impersonal physics determines everything that goes on, everything we think, decide and do, permitting us merely the illusion that we are in charge of our actions. The God-of-Cosmic-Power triumphs, and the God-of-Cosmic-Value turns out to be mere shadow-play, a mockery.

We have here a much more severe problem than the one tackled in chapter two. There, the task was to see how the experiential could be accommodated alongside the physical. The solution - I argued - is to appreciate that physics is only about that which determines how events unfold as they do, physics thus being silent about everything that does *not* determine how events unfold, and hence being silent about the *experiential* aspect of things. This solution concedes that the physical is solely and entirely responsible for the way things occur. It is at once clear that this solution, this line of thought, cannot solve our present problem. For what is at issue is *our* capacity to make things happen. We must be, it would seem, in direct competition with physics, with the God-of-Cosmic-Power. Physics cannot be solely responsible for what goes on. We must be responsible for at least some of what happens - namely what we do, what we create, even what we think, what we imagine and decide. Some power, some responsibility for making things happen, must be wrested from physics, it seems, if we are not to be merely the puppets of physics, our existence as acting beings an illusion.

Philosophers tend to formulate this as the problem of how (or whether) there can be free will if determinism holds. But this traditional formulation is inadequate in a number of respects.

First, the problem is generated, not by determinism but by physics - or rather by the truth of physicalism.[1] Physicalism is not at all the same thing as determinism. Determinism is both too broad, and too narrow. Determinism might be true, every event might be strictly determined by prior states of affairs, and yet physicalism might be false, and the universe might not be physically comprehensible. The true theory-of-everything of this universe, capable of predicting events, might be grotesquely complicated and disunified, and thus non-explanatory. On the other hand, physicalism may be true, the universe may well be physically comprehensible, and yet determinism may be false. The true physical theory-of-everything may be probabilistic, not deterministic – capable only of predicting events in probabilistic fashion. Determinism neither implies, nor is implied by, physicalism.

Second, the traditional formulation of the problem is inadequate in suggesting that what is at issue is free will. Much more than that is at stake. It is our very existence as acting persons that is under threat. We exist as persons only if we do things in this world, initiate actions. If the physical is solely and entirely responsible for everything that goes on, including of course everything going on in our brains, then we can be responsible, it seems, for nothing. We do not exist. Our existence is an illusion.[2]

---

[1] Physicalism is the doctrine that the universe is perfectly *physically* comprehensible, it being such that there is some yet-to-be-discovered, true, unified physical "theory of everything" (with N = 1). This is just the assertion that the God-of-Cosmic-Power exists.

[2] Traditionally, discussion about free will takes the form of a debate between compatibilists, who hold free will and determinism to be

In the end, the problem is simple and stark. Granted that the God-of-Cosmic-Power exists, and physics is responsible for everything that happens, there can be no room for us too. We can do nothing. Everything we seem to do is really the God-of-Cosmic-Power, physics, acting through us. We persons, and all other conscious and sentient beings, key embodiments of the God-of-Cosmic-Value, shrivel and die before the unchallengeable might of the God-of-Cosmic-Power.

In tackling this problem, it is a matter of profound importance whether we adopt the "internalist" theory of perception indicated in chapter three, and Cartesian dualism so closely associated with it, or the "externalist" theory, which permits us to hold that our inner experiences, our states of consciousness, are brain processes.[3] If we uphold Cartesian dualism, we have had it. Reconciling our

---

compatible, and incompatibilists, who hold they are not. Compatibilists argue that determinism does not imply compulsion, or lack of responsibility for one's actions or decisions, and thus does not imply there is no free will, and *indeterminism* can hardly enhance free will because it would introduce an element of chance or probability into human action. Incompatibilists argue, in opposition to this, that free will requires that there are genuinely open possible futures before us, between which we have the power to choose (not possible if determinism is true), and struggle to rebut the suggestion that indeterminism must *undermine* free will. For an excellent introduction to the debate, construed in these terms, see Kane (2005). For a lively defence of compatibilism see Dennett (1984). The best defence of incompatibilism known to me is Kane (1998). For a critical discussion see my (2001), pp. 151-154. For a defence of the view that the problem ought to be formulated in such a way that it is what science tells us about the world, or physicalism, that poses a threat to free will, not determinism, see my (2005a). In this chapter I concentrate on those issues that seem most relevant to the free will/physicalism problem rather than the traditional free will/determinism problem – although, of course, there is considerable overlap between these two formulations.

[3] The terms "internalism" and "externalism" are widely used in the philosophical literature to draw distinctions between a wide variety of views. The distinction I have in mind is the one drawn in chapter 3. It should not be identified with any distinction current in the philosophical literature: see note 12 of chapter three.

existence with the God-of-Cosmic-Power, with the universe being physically comprehensible, is impossible. In order for our conscious intentions to affect what goes on in our brains, and what our bodies do, it is necessary for these non-physical mental states or events to influence physical processes occurring in our brains, which means that some physical processes are not fully explicable physically. But if, on the other hand we hold the brain process theory, allied to externalism, there is at least a glimmer of hope. For then conscious intentions that initiate and guide our actions *are* themselves physical processes – and these physical processes can be a part of the cause of our actions without any physical laws being violated at all. There is just the faintest hint that "free will", and our existence as acting persons, might be compatible with physicalism.

In the end this vital point is extremely simple. If Cartesian dualism is true, then we – our conscious selves – are entirely distinct and separate from the physical universe. If my Cartesian conscious self is to be able to make my physical body act, then it must interact with physical processes going on in my body – physical processes going in my brain we may presume – thus violating the sole determination of physics, of the God-of-Cosmic-Power. But if Cartesian dualism is false, the brain process theory is true, and all my conscious states, intentions decisions to act are themselves *brain processes, physical processes*, then it becomes possible that my conscious intentions to act cause and guide my actions *without any physical laws being violated at all*, everything occurring in accordance with the as-yet-undiscovered physical theory of everything, T. In initiating action I have all the power at my command of the God-of-Cosmic-Power. I am a bit of the God-of-Cosmic-Power.

But there is still the mystery, of course, of how it can be possible for me to be in charge of my thoughts and deeds *even though the physical processes that are these thoughts and deeds unfold in a precisely determined way in accordance with the laws of physics*. I may be a bit of the God-of-Cosmic-Power, but nevertheless I am doomed, it seems, to think and act precisely as the God-of-Cosmic-Power dictates. How can I, and the God-of-Cosmic-Power *both*, simultaneously, be in charge, be responsible for what occurs? How is dual control possible?

In order to explore this question, let us assume that Cartesian dualism and the "internalist" theory of perception are false. We directly perceive things external to us, and our inner experiences are brain processes. Or rather, to state the last point more carefully, processes going on inside our heads – "head processes" as we may call them – have both a physical aspect and an experiential aspect. Head processes are neurological processes going on in our brains; but some of them are also conscious inner experiences, thoughts, feelings, desires, decisions to act, sensory experiences, imaginings – all the contents of our rich inner life. That head processes have an experiential aspect and a physical aspect is somewhat analogous to the way honeysuckle, let us say, has an experiential aspect – its colour, its smell – as well as a physical aspect. Or, to give another example, it is somewhat analogous to the way this very sentence has an experiential aspect – its meaning, what it asserts – and a physical aspect.

Purpose in a Physicalistic Universe

One point deserves to be appreciated straight away. There is no problem whatsoever in understanding how it is possible for there to be things able to pursue goals – purposive beings in other words – in a fully physicalistic universe, even one that is *deterministic*. The problem of how this is possible is solved by the feedback mechanism – the unit of control. Devices, from thermostats to guided missiles and robots, are able to pursue goals even though acting wholly in accordance with deterministic physical laws, because they incorporate feedback mechanisms in their bodily structure.

One of the simplest examples of a feedback mechanism is the thermostat. This consists of a heater, a thermometer, and a switch. The aim of the thermostat, let us suppose, is to keep the room at the temperature of 20^0 centigrade. It achieves this by switching the heater off if the temperature rises above 20^0 C, and switching it on if it falls below 20^0 C.

We can imagine this being done as follows. As the temperature of the room falls below 20^0 C, the metal strip, of which the thermometer is composed, shrinks in length and, as a result, makes contact with an electric circuit which turns the heater on. The room heats up as a result, the metal strip expands, and this breaks the electric circuit, which turns the heater off. The outcome is that the thermostat acts so as successfully to attain the goal of keeping the room more or less at 20^0 C.

We have here the most elementary kind of negative feedback mechanism in action conceivable – the atom of control. It can be elaborated in various ways. First, instead of the action being the discrete one of ON/OFF, the action may rather be to increase or decrease something smoothly, for example, to turn the heater up by degrees or down, or to steer a rocket to the left or to the right to a greater or lesser extent. The action of the primitive ON/OFF thermostat could be represented by an arrow which points at one or other of two positions, ON and OFF. The action of a purposive device which acts continuously can be represented by an arrow which points at some point on a line, and moves to the left or right smoothly, along the line.

This latter continuously operating feedback mechanism can be elaborated by increasing the number of dimensions of continuous variability, from one, to two, three, … to 10,000, to N, where N is any number equal to or greater than one. A guided missile might, for example, have a three dimensional feedback mechanism for control system, guiding the missile to change its direction upwards or downwards, to the left or right, and to change its speed to go faster or slower. More complex control systems might consist of a number, M, of distinct control systems, each performing distinct control tasks, and a master control system which decides which of these M control systems is to operate at any given moment. One can even imagine a control system consisting of a hierarchy of control systems. At the base of the hierarchy, there are a vast number of control systems. As one goes up the hierarchy, the number of control systems decreases until, at the top, there is just one master control system activating and controlling all the others. It is possible that the brains of animals and humans are hierarchical control systems of this type. I formulate the intention to go down stairs and put the kettle on for a cup of tea – and think no more of the matter, acting more or less on autopilot, my thoughts elsewhere. This is the master control system activating subsidiary control systems to do its bidding. Control systems lower down in the hierarchy (in my brain) then guide my body to get up, walk out of the room, go downstairs into the kitchen, pick up the kettle, fill it from the tap, put it on the stove and light it. These midway control systems activate control systems still lower down in the hierarchy (in my brain) to move my legs, arms and hands appropriately so that these actions are performed; and further control systems, still lower down in the hierarchy, of which I am entirely unaware, control individual muscles in my legs, arms and hands so that the sequence of actions is performed. And this hierarchical structure of control, which might be true of me, might also be true of a monkey swinging from branch to branch in the forest, or of a fox out hunting for rabbits or mice.

This hierarchical hypothesis may, or may not, be true. The crucial point is that there is no problem whatsoever in understanding in principle how purposive action is possible in a physicalistic universe – even complex purposive action of the kind that mammals, and even humans, perform in life. We are still profoundly ignorant of the way mammalian brains, and human brains, produce the complex purposive actions that they do produce. But purposive action in a physicalistic universe does not in itself pose a problem of principle. The key to the solution of this problem is the feedback mechanism. It is worth noting that, for the feedback mechanism to work properly, to produce the purposive action that it is designed to produce, it is essential that deterministic physical laws, on which the feedback mechanism depends, continue to operate. To return to the atom of control, the thermostat, if, as the temperature falls, the metal strip began to *expand* (and not shrink), this would play havoc with the capacity of the thermostat to achieve its goal and control the temperature of the room. Far from purposive action, produced by control systems (based on elaborations of the feedback mechanism) being *in conflict* with physicalism, it is all the other way round: purposive action of this type actually requires physicalism to be true – at least as far as the workings of the control system of the purposive actor is concerned.[4]

Free Will in a Physicalistic Universe

What I have said in the previous section solves the problem:-
(1) How is purposive action possible in a physicalistic universe?
It does not, however, solve the problem:-
(2) How is free will possible in a physicalistic universe?

The hypothesis we are exploring is that we *are* purposive beings of the kind that we have been considering, but with the utterly amazing additional features that we are sentient and conscious, our conscious intentions more often than not guiding our actions, we ourselves guiding the flow of our consciousness (at least to some extent). Dual control is possible because a part of the *physical* states of affairs that determine what we do are brain states which

---

[4] The notion of purposive action has been bedevilled by intellectual history, in particular by the lingering influence of Aristotle. It was Aristotle's view that the physical universe is to be explained and understood in terms of the notion of purpose. Physical things have an inherent purpose built into them, and that is the explanation as to why they behave as they do. Stones fall because they seek their natural resting place, the centre of the earth. It was one of Galileo's great battles to oppose this purposive metaphysics with the idea that "the book of nature is written in the language of mathematics" – an early version of physicalism. One outcome of this battle has been that scientists after Galileo have tended to take for granted that the Aristotelian notion of purposiveness is incompatible with modern physics, incompatible indeed with modern science. Even biologists have made this assumption, holding that it cannot be correct to attribute purpose to animals, to living things. The all-important point is to note that there are two notions, the Aristotelian, incompatibilist one, and a post feedback-mechanism, compatibilist one. The first holds that purposiveness is incompatible with physicalism, whereas the second holds that it is compatible. Both notions are legitimate. The crucial question, however, is: Is there anything that is purposive in the Aristotelian, incompatibilist sense? Those who hold that we have free will, and free will is incompatible with physicalism, in effect hold that we are purposive beings in the Aristotelian sense. My view, of course, is that we are *not* Aristotelian beings, and there are none in existence.

are also *conscious* states – our decisions to act. In deciding to act we have behind us all the might of the God-of-Cosmic-Power. We *are* a bit of the God-of-Cosmic-Power. Our amazingly structured brains and bodies enable us to exploit the bit of the God-of-Cosmic-Power that we are to do our bidding, do what we wish to be done, so that It becomes our servant, not our master. Our decision to act *is* that bit of the God-of-Cosmic-Power that, in the context of our brain, body and environment, produces and guides the actions that we have decided to perform.

Let us call this suggestion the *dual control* hypothesis.[5]

The *Compatibilist/Incompatibilist* Debate

I now subject this suggestion to a debate between an *Incompatibilist* and a *Combatibilist*.[6]

Incompatibilist: But how can we really be in charge of, and responsible for, our thoughts, decisions and actions if all these unfold precisely in accordance with iron physical law? It may feel as if we are in charge, free to choose what we think, decide and do, but really this is all prescribed with absolute precision by physics. We are held in the vice-like grip of physicalism, not a twitch or momentary thought possible that is not predetermined by physics.

---

[5] "Dual control" is perhaps somewhat of a misnomer. The God-of-Cosmic-Power may, with variable instantaneous states of affairs, *determine* what happens next, but It does not really *control* anything. Only feedback mechanisms, control systems, do that. Still, one could think of the God-of-Cosmic-Power as having a hand in control whenever a control system does exist, via the particular, persisting and variable states of the control system. "Dual control" could be thought of as referring to the brain conceived of purely as a physical system, and the brain conceived of as the seat of consciousness. Nevertheless, "dual determination" might be a better title for the view than "dual control".

[6] *Compatibilist* and *Incompatibilist*, here, differ as to whether or not free will is compatible with *physicalism*, and not, as in note 2, with *determinism*. Physicalism may be either deterministic or probabilistic (the former being a special case of the latter). My own view, as I made clear in the last section of chapter 5, is that we should interpret quantum theory as trying to tell us that nature is probabilistic, the interpretative problems of the theory arising from the failure to take probabilism seriously: see my (1988); (1994); (2004b); and (2010b).

Compatibilist: But this just ignores that, even though physicalism is true, nevertheless a person's brain, body and environment may well be such that conscious decisions to act really do produce and control the actions that are performed – these conscious decisions being neurological processes that produce and control the person's body to do what the person has decided to do. And likewise, in pondering, day-dreaming, imaginatively exploring some issue, or thinking about what to do, the person's brain, body and environment may well be such that the person is indeed in control of what he or she is thinking, the state of mind being a brain state which produces and controls neurological processes that are the ponderings, day-dreams, imaginings or thoughts of the person in question.

In the Autumn of 2007, at the time of writing, Gordon Brown, then the prime minister of Britain, agonized about whether or not he should call a general election. Here, let us suppose, are his thoughts. Winning a general election would give his premiership a legitimacy it does not otherwise quite have [since he took over without one when Tony Blair (the previous prime minister) resigned]. Unofficially, the Labour party has prepared for a general election, and all the news media have declared that a general election will be called – strong reasons for calling an election. On the other hand, the polls are not too good, and he might lose, after having been prime minister for only a few months. So far he has managed to avoid saying anything in public to indicate that he is even contemplating calling an election. Weighing up the pros and cons, he decides not to make the call for an election (the decision he did, in fact, make).

It is possible, of course, that Gordon Brown is acting compulsively, driven by an irresistible fear of failure not to risk an election. But let us suppose this is not the case. He thinks, feels, ponders, decides and acts freely. Now suppose experiential physicalism is true.[7] Gordon's brain is such that his conscious thoughts, feelings, ponderings and decisions are neurological

---

[7] Experiential physicalism, as I explained in chapter 5, section (4), is physicalism plus the thesis that the experiential exists *in addition* to the physical, and cannot even in principle be reduced to the physical.

processes – ultimately physical processes – that guide and control other conscious states in just the right way for them to be Gordon's freely chosen thoughts, conscious states of mind, decisions, even though they all occur precisely in accordance with physical law. Why does this supposition of experiential physicalism deprive Gordon of free will? What is it about free will that makes it impossible to be compatible with experiential physicalism?

Incompatibilist: Physicalism makes it impossible for Gordon to be thinking and acting freely because it makes it impossible for him to think and do otherwise than what he did think and do. An essential requirement for free will is that, even though everything is precisely the same, one could have done otherwise. Physicalism makes this impossible.

Compatibilist: First, if this requirement is conceded, it only renders free will incompatible with *deterministic* physicalism, not with *probabilistic* physicalism. But second, is it sensible in any case to demand, for free will, that even though *everything* is the same, one might have done otherwise? If this condition is satisfied, does it not rather *undermine* free will instead of strengthen it? It would mean that our decisions to act are sometimes subverted by the outcome being other than what we had intended. This hardly seems to support, to strengthen free will.

There are those occasions, it is true, when we act in a wholly spontaneous fashion, on impulse, without prior deliberation, even in a way which takes us by surprise, and yet – we may hold – at such moments we seem to be at our freest, most ourselves, least restricted by conforming to conventions which constrain our freedom. Such spontaneous, impulsive action may even be associated with creativity of the highest order. It may feel as if we might have performed any number of different actions, this range of possibilities being open to us having everything to do with the fact that we acted with a high degree of free will.

It may be conceded that spontaneity and impulsiveness are integral to free action. He who can never act spontaneously, but must always deliberate consciously before acting, never allows himself to act in an unconstrained way. Such a person is an obsessive, compulsive deliberator and, to that extent, somewhat lacking in free will. What matters is that we are able to ponder and deliberate when it is necessary, when action becomes problematic – as it undoubtedly is for Gordon Brown in deciding whether or not to call an election. None of this means, however, that when we act spontaneously it really is the case that, even if everything had been precisely the same, we might have done otherwise. Spontaneity is perfectly compatible with (experiential) physicalism, and compatible, too, with the fact that we *could not have done otherwise* if everything had been exactly the same.

An explanation can, however, be given as to why it might seem plausible to hold that this requirement for free will [8] is a reasonable one to make. Suppose one upholds some version of dualism. The conscious mind is distinct from the brain. This means that, for free will, the conscious mind must be able to effect changes in the physical brain. It must be the case, in other words, on at least some occasions that, given a definite *physical* state of the brain (and the rest of the universe), the conscious mind can get the brain, and hence the body, to do a number of different things, depending on what it decides it wants to do. If this is never possible, the conscious mind cannot affect the brain in any way whatsoever – and there can be no free will. In short, given dualism, it is inevitable that one demands for free will that one could have done otherwise even if everything *physical* had been precisely the same.

But if dualism is rejected, experiential physicalism and the externalist account of perception is accepted, and inner experiences are held to be brain processes, all this looks very different. To demand that, for free will, it must be the case that one could have done otherwise *even though everything physical is precisely the same*, is to demand that one could have done otherwise *even though one's entire mental state was precisely the same*. In demanding everything physical is the same one thereby demands that everything mental is the same as well (since, granted the brain process theory, it is not possible for the brain and the rest of the universe to be the same but the mind to be different). But it is not reasonable to require, for free will, that we could have done otherwise *even though one's entire mental state is precisely the*

---

[8] The requirement, that is, that we could have done otherwise even though absolutely everything is the same.

211

same. This makes our actions – or perhaps our decision-making – unpredictable and, to that extent, out of our control. As I have already said, that would *undermine* rather than *support* free will.[9]

Granted the truth of the conjecture that our conscious inner processes are brain processes, what we require for free will is that, if things had been different in a range of possible ways, we would have been able to act in appropriately different ways so as to achieve our goals. Consider Gordon Brown again. If he was pathologically incapable of calling an election, whatever the circumstances, then we would have to say that he was, to that extent, acting compulsively, and not freely. For free will we require that Gordon would have acted differently in different circumstances. We require, for example, that he would have called an election if the Labour party had been higher in the polls, or if he had known just how dire his unpopularity would become if he did not call an election (as it did become). He might have called an election if it had been leaked to the press that he had ordered the Labour party to prepare for an election, since then it would have seemed wholly unacceptable for him to change his mind.

Acting differently, in appropriate ways, if things had been different is of the essence of the dual control hypothesis that we are considering. For this is basic to the very idea of a purposive being, whether it be robot, insect, bacterium, plant or animal.[10] If a thing is to pursue a goal, then it must be able, in its given environment, to vary its actions appropriately, as the environment varies, so that the goal is achieved in a range of circumstances. The very idea of purpose, of pursuing a goal, disappears if this is not the case. The goal-pursuing thing must be able to respond appropriately to some

---

[9] This is the case even if dualism is true. If everything is the same, both physical state of the brain and environment, and mental state of the mind, but one could nevertheless have done otherwise, this means that one's mental states do *not* always determine one's actions. The link between intentions and actions is sometimes at most probabilistic. This does not enhance free will.

[10] All living things are purposive (a point to be discussed in the next chapter). Darwinian theory tells us that they have, as their basic goals, survival and reproductive success. Plants pursue these goals, in the main, by means of growth.

range of different possible circumstances so as to attain its goal in all of them, even if, outside this range, the thing fails, and the goal is not achieved. It is just this appropriately varied action, responding to different circumstances so that the goal is achieved in all of them, which the feedback mechanism contrives to produce. In the case of a guided missile, for example, the feedback mechanism is able to guide the missile back onto its predetermined course when it is blown off course by winds – of any direction and a range of strengths. What applies to guided missiles, robots, insects and plants applies to sentient and conscious dual control beings as well, mammals and people.

Incompatibilist: Here is a rather different consideration. If deterministic physicalism is true, then the entire history of humanity was determined by the state of the universe just after the big bang. Given an instantaneous state of the universe at or soon after the big bang, and given the basic physical laws, all future states of the universe follow necessarily. But we can't alter the past, we can't alter the laws of nature, and we can't alter what follows necessarily from these two either. Hence, we cannot alter the future now. We are devoid of free will.[11]

Compatibilist: It is the third step in this argument that must be challenged. We can't alter (1) the past, and we can't alter (2) the laws of nature, but we can alter what follows necessarily from (1) and (2) if we are a part of what follows, in that we have the power, now, in a compatibilist sense, to alter what will happen in the future. Having the power to alter what will happen in the future, in a compatibilist sense, means having the power to do otherwise, in relevant ways, if the circumstances had been, in various ways, somewhat different. Determinism does not deprive us of this compatibilist power to do otherwise. Even though we have no power to alter the past or the laws of nature (even in the compatibilist sense) we do have the power, given determinism, to alter (in a compatibilist sense) what would follow necessarily from a complete specification of the distant past plus the laws of nature – namely, some of what lies in our future.

---

[11] For a more detailed version of this argument see van Inwagen (1986).

213

Admittedly, *orthodox incompatibilists* (those who hold free will and determinism are incompatible) find this compatibilist power "to do otherwise" insufficient for genuine free will. They hold that free will requires that two or more possible futures lie before us, and we have the power to decide which one occurs. But this faces the difficulty, already mentioned, that *probabilistic* physicalism, far from enhancing free will, can only serve to undermine it in that, if probabilistic jumps occur during the process of decision making, or during the execution of decisions already made, a decision is formed other than what we intended, or an action executed other than what we decided – none of which strengthens free will. This is the problem that faces orthodox incompatibilism: how can moving from *deterministic* to *probabilistic* physicalism do anything other than undermine free will?

There is a possible way to solve this problem, to some extent at least, but it is a line of thought that seems to be of little interest to orthodox incompatibilists. It involves concentrating on probabilistic events that occur *outside* rather than *inside* people. The view might be called *environmental probabilism*. In people's brains, let us suppose, probabilistic events, if they occur, do so in such a way that free will is not thereby undermined. External to people's bodies probabilistic events occur which are such that distinct macroscopic states of affairs result, depending on which way each probabilistic event went. People would have to respond to different circumstances thrown up by these probabilistic events. The unpredictability in principle of some aspects of the environment creates in general, we may suppose, no special problems for the successful exercise of free will. The probabilistic events are, in practice, indistinguishable from deterministic but unpredictable events – unpredictable, like the weather, because of lack of information and predictive power. We have here, then, probabilistic physicalism but no undermining of free will. The universe is such that (1) the state of the universe just after the big bang, and (2) the basic laws of nature do *not* determine human history. The future might be determined *probabilistically* in such a weak fashion that, given (1) and (2), our present is but one of infinitely many alternative possibilities, most, perhaps, having no human history at all. There is a genuinely open future, only probabilistically, and perhaps very weakly, determined; and yet this move from determinism to probabilism involves no loss of free will. The net result may be regarded as a strengthening of the free will that is possible in such a version of physicalism.

Incompatibilist: There is another way to strengthen the kind of free will that is possible in moving from deterministic to probabilistic physicalism. We live, let us suppose, in a probabilistic universe. Gordon Brown has to decide between going ahead with the general election or calling it off and, just before he makes the decision, the physical state of the universe is such that each option has a probability of a half. Gordon Brown makes his decision, but the decision is not made in a manner analogous to tossing a coin. On the contrary, cogent reasons emerge in Brown's mind as to why calling an election is not sensible: the way he makes the decision makes human sense, and is a free action, a free decision. All this is reflected in what goes on in Brown's brain. Nevertheless, from the standpoint of physics, before the decision is made there is merely a probability of one half that the election will be declared, a probability of a half that it will be called off. At the human level, what went on satisfies all the conditions required for a freely chosen decision, but at the level of physics there is merely the probabilistic prediction. In these circumstances, probabilism enhances free will.[12]

Compatibilist: What is being proposed here, I take it, is that probabilism enhances free will in the sense that there is a kind of free will, incompatible with *deterministic* physicalism, which is nevertheless compatible with *probabilistic* physicalism, where the probabilistic transitions occur, not in the environment, but in the brain.[13]

I find this suggestion implausible. As I have already said, some brain events being probabilistic – as opposed to all being

---

[12] For expositions of arguments along these lines, see Kane (1998), especially pp. 72-101 and 124-195. For criticism see my (2001), pp. 153-154. See also Hodgson (2005) and my criticism (2005b).

[13] Kane defends a version of this view. Hodgson, however, defends incompatibilism, and the reality of free will, whether physicalism is deterministic or probabilistic. See, in both cases, works referred to in the previous note.

deterministic – can only, so it would seem, *undermine* free will. For it would mean that sometimes decisions to act do not result in the action decided on, because a probabilistic event intervenes. Or it would mean that our meaningful flow of mental life is disrupted by probabilistic events. Even if the idea is that a person, choosing between two actions, is freer if the choice is determined only probabilistically and not deterministically, this still suffers from the defect that it reduces decision-making to the equivalent of tossing a coin (which would not ordinarily be thought to amount to an exercise of free will).

The suggestion may be, however, that probabilism allows for states of mind (to which correspond states of the brain) to determine subsequent mental states, decisions to act, or actions, in a way which is more definite and specific than that determined by the underlying probabilistic physics. Whereas deterministic physicalism leaves no room for this kind of psychological determination to operate in addition to the physics, probabilistic physicalism does provide room for it to happen, without basic physical laws being violated.

But this suggestion seems implausible for two reasons.[14] First, the suggestion requires brain states to determine subsequent brain states (or bodily actions) in a way which is entirely in addition to what is determined by fundamental (probabilistic) physics. Somehow, evolution has led to new physics – new physical laws – which operate, however, only in connection with conscious brains (or perhaps the brains of living things more generally). This is hardly plausible when viewed from a scientific standpoint. It is a version of vitalism – the discredited doctrine that special physical laws operate in connection with life. Second, it is difficult to see how this psychological or personal determination of subsequent states of affairs could operate in a way which is compatible with

---

[14] It might be objected that this view threatens to undermine incompatibilism by introducing a new kind of psychological determinism. But such an objection would amount to little more than a cheap debating point. What the incompatibilist has in mind, here, is not psychological laws constraining how mental states evolve, but rather intentions, decisions to act, reliably producing what is intended or what has been decided.

216

the basic probabilistic physics. If mental states, and thus brain states, can determine subsequent mental states (brain states) repeatedly in a more definite way than basic probabilistic physics can then, so it would seem, the basic physical laws must be violated. No doubt it would be all but impossible to demonstrate this violation experimentally, because of the extraordinary complexity of the brain. But it is difficult to see how complex brain states can determine what subsequent brain states occur in a way which is wholly in addition to what basic probabilistic physics can determine, without basic physics being violated. The move from deterministic to probabilistic physics, in short, does not seem to make it any easier to have mental determination that is wholly in addition to physical determination and that is, at the same time, compatible with fundamental physical law.

Incompatibilist: If, whenever we act, we could not have done otherwise even though everything had been the same, as deterministic physicalism implies, then there can be no such thing as *choice*. Choosing involves deciding which of two or more possible actions to take. This in turn requires that these options are real options, which might indeed have been selected. But if we could not have acted in a way other than the way we did act, then choice, alternative options, is an illusion. Choice is an illusion. And that means that free will is an illusion, for free will requires choice.

Compatibilist: It is not at all clear that having a genuine choice does require that many outcomes are possible, any one of which could have happened given the state of affairs before the choice is made. For this may make choice an arbitrary matter, not a matter of goals, reasons, desires. If, before Gordon Brown makes his choice, the two possibilities are equally balanced and it is a purely probabilistic matter which is made, equivalent to tossing a coin, this does not seem to amount to choice in any very significant sense (as I have already remarked). What matters, for choice to be an authentic exercise of free will, is that there are reasons for the choice, some kind of personal explanation for it, couched in terms of the goals, desires, feelings, personality, context, beliefs, hopes and fears of the person in question. The all important point, from the standpoint of free choice, is that the person in question would have chosen differently, in appropriate ways, if the goals, desires, feelings, context, beliefs, hopes and fears had been different. For Gordon Brown's choice to be a genuine choice, an expression of free will, what is required is that his choice would have been different, in appropriate ways, if the situation had been different.

Incompatibilist: This conception of choice makes humans no freer than animals.

Compatibilist: Human choice has undoubtedly evolved out of animal choice. Animals often act in contexts of extreme conflict. A mouse must leave the safety of its burrow to forage for food, but in doing so risks its life from preying owls and foxes. The mouse must persistently decide whether it is safe to keep snuffling around for food, or whether the rustling this entails risks death, and it must freeze, or scuttle back to safety. The choice between two radically different lines of action that lie before the mouse is far more extreme and agonizing than choices that most of us humans have to make for most of our lives. Darwinian evolution designs the brains of mice, and of other mammals, to be good at making such choices, good at learning how to make such choices. To put it brutally, those mice that are not good at making such choices die (and thus fail to breed), falling victim either to predators or to starvation. Choosing well is vital for survival and reproductive success.

Our human capacity to choose has undoubtedly evolved from earlier mammalian capacity to choose well. But our human capacity to choose is massively enriched and enhanced by the massively enriched range of choices that lie before us – if we are fortunate – and our massively enriched human consciousness, founded on imagination and language (to be discussed in the next chapter). Human choosing has evolved from animal choosing, but is such a profound enrichment of animal choosing that free will may well seem to apply only to humans, and not to animals.

In both animal and human cases, however, the kind of choosing we have just considered is entirely compatible with physicalism, even deterministic physicalism. What matters, for free choice in this compatibilist sense, is the capacity to act differently in appropriate ways (so as to achieve desirable aims) if the circumstances had been different in a variety of ways.

Incompatibilist: If a person is to have free will, it must be the case that that person's authentic self initiates and controls the actions that that person performs. (It must be the *authentic* self, because if an inauthentic self is in control, a product, perhaps, of brainwashing, hypnosis or indoctrination, we should hold that the person is not in charge, not acting freely, or at least acting with severely diminished freedom.) But if physicalism is true, this cannot be the case. It is purely *physical states of affairs*, plus *physical laws*, which determine what occurs, not the person's authentic self at all. Free will is impossible.

Combatibilist: What is the self? Is it reasonable to hold, for the sake of this debate, that the self is that which enables the person to be what he or she most distinctively is, prompting or guiding the person to act in his or her characteristic ways, compounded of basic desires, fears, hopes, goals, a reservoir of basic memories, skills, knowledge, beliefs, values, temperament, likes and dislikes?

Incompatibilist: For the sake of the discussion, let it be conceded that the self is something that is roughly along these lines.

Compatibilist: Excellent. For, granted that we conceive of the self along some such lines, and granted the brain process theory of the mind, we can hold that the self *is* a basic part of the control aspect of the brain – that persisting part of the brain crucially involved in guiding the person to be what he or she most characteristically is, do the kind of things he or she does. According to the dual control theory, the self, so construed, does really guide and control what the person thinks, imagines and does. Experiential aspects of the self *are* control aspects of the brain which, in turn, *are* physical states and processes of the brain which physicalism itself declares to play a crucial role in determining what occurs. The task of distinguishing the *authentic* from the *inauthentic* self poses no special problem for this view. It may be that the distinction ultimately has to do with values, in that it depends on what is *of most value* in the person's character and life.

Incompatibilist: According to the brain process theory and dual control hypothesis, our conscious inner life is made up of the *experiential* aspects of brain processes. When we decide to do something, and we initiate the action, what we know about what goes on inside our head is the experiential aspect of the decision.

But it is the *physical*, not the *experiential* aspect of the brain process which plays a role in producing and guiding the action. The experiential aspect is causally impotent. It plays no role whatsoever. Thus our inner conscious life has no role whatsoever in initiating and guiding action.[15] It is at most merely a wholly impotent witness to what goes on. Hence, if experiential physicalism is true, there can be no free will.

Compatibilist: But the decision to act, as experienced by the person in question, *is* the brain process, the physical state of affairs, that plays a crucial role in initiating and guiding the action performed. The experiential decision is contingently identical to the physical process going on in the brain just as the morning star is contingently identical to the evening star, both, as a matter of fact, being the planet Venus.[16]

Incompatibilist: Even if this is the case, nevertheless it is still only the *experiential*, the *mental*, aspect of brain processes that the person knows about, and this aspect plays no role in initiating and guiding the actions performed. Hence, what the person knows about – in a sense, what the person *is* – plays no role in initiating and guiding action. The person is condemned to impotency.

Compatibilist: This criticism is valid when directed against a well-known view in the history of thought about the mind-brain problem, namely *epiphenomenalism*. It is not valid, however, when directed against the view being defended here (brain process theory plus dual control hypothesis). Epiphenomenalism is a version, a modification, of Cartesian dualism. According to epiphenomenalism, the mind, consciousness, is distinct from the brain. Brain processes cause distinct mental phenomena to occur in the conscious mind, but these mental phenomena have no causal impact on the brain whatsoever. The causal influence is in one

---

[15] There is considerable discussion of this issue in the philosophical literature: see, for example, Robb and Heil (2008), and references therein cited.

[16] A famous – or infamous – argument by Saul Kripke claims to show that contingent identity of the kind involved here (with "rigid designators") is not possible: see Kripke (1981). Elsewhere, I have demolished this argument; or rather, shown that there really is no argument: see my (2001), Appendix Two.

direction only, from brain to mind (whereas Cartesian dualism asserts that the causal influence goes in both directions).

Your criticisms are entirely valid when directed against epiphenomenalism: conscious decisions to act can, given this view, play no role whatsoever in producing and controlling the action decided upon. They are not valid, however, when directed against the very different view of the dual control hypothesis. Granted this view, conscious decisions to act *are* (contingently identical to) brain processes that play a crucial role in producing and controlling the actions decided upon – not, as epiphenomenalism would have it, distinct from such brain processes, and having no causal impact on these brain processes.

Furthermore, in being aware of the experiential aspect of a decision to act – get up from one's chair and greet a friend, let us say – the person in question is aware of precisely that aspect of the brain process that the person needs to be aware of in order to be in control of his actions, namely the *control aspect* of the brain process. We are ordinarily blithely ignorant of the *physical* aspects of our inner experiences. That it is why it is such a shock when we first learn that our inner experiences, our thoughts, feelings, perceptions, imaginings, states of awareness, are all *neurological processes* – complex patterns of neurons firing, waves of potassium and sodium ions being exchanged through the semi-permeable membranes of neurons plus parcels of chemicals leaping the gaps of synaptic junctions. How can this delicious scent of honeysuckle, this vision of leaves and sun and blue sky, this murmuring voice, these feelings of joy, sorrow, weariness, anxiety, these aches of longing and desire, this whole world of rich, subtle, ever-varying awareness, all actually *be* intricate waves of potassium and sodium ions chasing down complex arrays of neurons in the brain? So absurd does this seem that it is no wonder there is this strong tendency to believe that our inner experiences are quite distinct from anything going on in the brain.

But what we ordinarily know about what goes in our brain is, as I have said, what we need to know about to be in control, namely the *control aspects* of brain processes. A person sees a chair in front of him and decides to sit down. In seeing the chair, the person is in effect aware of a brain process going on in his brain.

He is aware, first and foremost, of the chair, but secondarily, of his perception of the chair. But what the person is aware of is not the neurological or physical aspects of this process – the neurons, the potassium and sodium ions, etc. – but the *control* aspect, that aspect relevant to action, to the successful realization of goals. The eye and brain of the person function in such a way that the brain process in question contains information about the nature of the object before him: it is a chair. It is this vital control feature of the brain process that the person becomes aware of (in an incidental way) in seeing the chair. The decision to sit down is another complex neurological process. In being aware of the experiential features of this process, the person knows nothing of the neurological or physical aspects of the process, but instead knows what the *control* aspect is: this, and just this, neurological process, occurring in this brain, will lead to the body moving, turning, and, as arms reach out for the arms of the chair, the sinking down so as to be sitting in the chair. We all have, in short, quite incredibly sophisticated knowledge of the control features of neurological processes going on in our brains. We would not be able to perform the actions that we do perform were this not the case. This wonderfully intricate and sophisticated knowledge that we possess of the control aspects of our brain processes is something that we have to *learn*. Young babies do not have it. One can sense, sometimes, that a baby does not know how to grab a foot to suck a toe: the arms and legs thrash about in frustration. We *learn* about the control aspects of processes that go on in our brains, and as we learn we no doubt also *create*; we no doubt *develop* our brains to have increasingly sophisticated control structures and functions. Our brains come to reflect what we have experienced and struggled to do in the past. We make our brains up, to some extent, as we live.

In short, what we know about neurological processes going on in our brains is just what we need to know in order to see, hear, touch, speak and *act*, namely the control aspect of these processes. It would be a disaster if we knew about the *neurological* aspect, and an even greater disaster if we knew about the *physical* aspect. We would be paralysed with irrelevant information overload, lost for ever among our neurons and synaptic junctions. Our inner experiences are, nevertheless, neurological processes, physical processes. In being aware of the control aspect of these processes, we are aware of what we need to be aware of in order to be able to act more or less successfully in the world, and it is these processes that play a crucial role in initiating and guiding what we do. If this were not the case, we would not be aware of the control aspect after all.

None of this means, of course, that we have anything like a *complete* or *infallible* knowledge of the control aspects of processes going on in the brain. People misconstrue what they see. They deceive themselves about their motives, desires, feelings and beliefs. They misconstrue what it is they are doing, and why they are doing it. And no doubt much that goes on in the brain has no experiential aspect at all, but instead makes the occurrence of other brain processes possible which do have experiential aspects.[17] We all have an incredibly detailed and sophisticated knowledge of the control aspects of neurological processes going on in our brains, but some of us have a more detailed, sophisticated and accurate knowledge than others.

An important, additional point needs to be appreciated. It is not just that we have incredibly sophisticated knowledge of the control aspects of processes going on inside our heads. In addition, we have an incredibly sophisticated capacity for creating head processes subtly and brilliantly designed to have just those control

---

[17] A basic task for psycho-neurology is to discover how to correlate control or functional aspects of brain processes with neurological aspects, it being especially important to pin-point those functional aspects that are *experiential* or *conscious*. A basic task, in short, is to locate consciousness in the brain. For a discussion of this problem, and a proposed outline of a solution, see my (2001), ch. 8. I there suggest that consciousness is to be located in the limbic system plus whatever other part of the brain the limbic system is in strong, two-way interaction with at that moment. This suggestion does justice to the idea that consciousness should have some kind of fixed location in the brain (centrally placed, if possible) and at the same time does justice to the idea that consciousness flits about the brain as we become aware of, and think about, different things – the visual, the auditory, motor control, abstract thought, remembering past events.

features required to enable us to do what we have decided to do. I decide to get up and greet a friend. This decision *is* the neurological process going on in my brain responsible for activating my brain to control my body to get up, walk towards my friend and greet him. Even though I ordinarily know nothing about the neurological or physical aspects of the brain state that guides my actions, I nevertheless have this extraordinary capacity to create just the neurological process required to execute my intention, the neurological process in question being conceived of, by me, in control terms. I have the power, the capacity, to do endlessly many different things, and thus the capacity to create endlessly many different neurological processes or states in my brain all precisely and brilliantly crafted, in control terms, to do the endlessly many different things that I can do.[18]

This view has the great merit of doing justice to two apparently mutually contradictory features of the situation simultaneously. It does justice to our ordinary complete ignorance about processes going on inside our head while at the same time doing justice to our highly detailed and sophisticated knowledge of these very same processes. We ordinarily know next to nothing about head processes conceived of as physical or neurological processes, but know a great deal about these same head processes conceived of in control terms. It is, in part, this duality of absolute ignorance and detailed knowledge that leads some to uphold dualism: there are two kinds of processes, physical processes (of which we are ordinarily ignorant) and mind processes (about which we know a great deal). There is, however, another option: there is just one kind of process, the head process: this may be conceived of and specified, either as a physical (or neurological) process, or as a control process. That there are these two kinds of specifications does not mean that there are two kinds of process.

Incompatibilist: But in the end it is not this so-called "*control* aspect" of brain states or processes that initiates and controls actions performed, but the *physical* aspect of these states or

---

[18] This knowledge of, and control over, the brain by the brain, is a key feature that psycho-neurology will need to depict and comprehend in order to understand the workings of the mammalian and human brain.

processes, and of that aspect of processes going on inside our heads we ordinarily know nothing. Thus, what really determines what we do – the physical features of what goes on inside our heads (plus of course physical features of our bodies and environment) – is something we ordinarily know nothing about. That in turn means that *we* are not in charge. *We* cannot be responsible for what we do. If experiential physicalism is true, there can be no free will.

Compatibilist: The key step we need to take, I think, to see that experiential physicalism *does* make free will possible is to specify, in general terms, the physical structure a brain would have to have to possess what we may call "perfect free will". It goes like this. A conscious brain with perfect free will has a neurological, and ultimately *physical* structure such that the physical aspect of any conscious head process behaves in just the way required for the conscious content to *be* what it purports to be, or *do* what it purports to do, acting so as to promote conscious free will. In such a brain, the neurological process that is the conscious perception of a rose has been caused to occur by a rose-like image being projected onto the retina of the eye; furthermore, the neurological process is such that it leads to other physical processes occurring in the brain, which constitute such things as awareness of the rose, knowledge of the rose, and even, perhaps, if it is wished, the conscious decision to move forward and smell the rose. This conscious decision, in turn, is such that its physical aspect causes muscles to contract in precisely the way required for the decision to be enacted, in just the way the decision determined. If circumstances had been different, in a range of possible ways, then conscious processes, and the actions they control, would also have been different in precisely the ways required for the person consciously and freely to attain her goals, the same or modified as the occasion would have required. In the conscious brain with perfect free will, in short, physical processes occur and evolve in precisely the way required by free, conscious action. The physical aspect of a conscious thought, perception, desire, feeling, or decision to act, is caused, evolves, and has causal effects precisely as required by the perception, feeling or decision that it also *is*. Physical processes in the brain conform to what consciousness

225

requires, not because consciousness intervenes and bends a law of nature or two, but because the particular, remarkable architecture of the brain arranges for this outcome to occur. Granted physicalism, two factors determine what occurs: (1) the laws of nature, and (2) particular, instantaneous states of affairs. It is the particular, instant by instant – and highly remarkable – states of the perfectly free conscious brain that ensure that physical processes, evolving in the brain in accordance with the laws of nature, do so in such a way as to serve the interests of conscious free control. Thus, in the brain with perfect free will, physics is enslaved to consciousness, to the authentic self, and the God-of-Cosmic-Power is the slave of the God-of-Cosmic-Value. (Although, for the latter, we require in addition, perhaps, that the brain with perfect free will *also* has perfect wisdom.)

Given that a conscious brain satisfies this remarkable requirement of having perfect free will, it makes perfect sense to say that consciousness is in control, conscious decisions to act initiating and controlling the actions decided upon. In the brain with perfect free will physics, we might say, is the means by which consciousness exercises its control.

Our human, partial free will exists because our brains satisfy partially, but no doubt rarely fully, this requirement of perfect free will.

For my argument to succeed, it should be noted, all we require is that conscious brains with perfect free will are *possible* granted experiential physicalism. And this we have. There is nothing in physicalism as such which precludes the possibility of the perfectly free brain, as I have characterized it.

Incompatibilist: But the conscious brain with "perfect free will" as you have described it, has no free will at all because everything is precisely determined by physics.

Compatibilist: But don't you see? What I have just said wipes out any reason you may have for holding that, granted experiential physicalism, physics must over-ride consciousness. In the brain with perfect free will, physics is wholly commandeered to serve the interests of conscious free action, and does not obstruct the freedom-producing actions of consciousness at all. What I have given is an account of how consciousness can dominate, guide and control physics in the brain within the framework of experiential physicalism. There is not a whiff, here, of the impotence of epiphenomenalism.

Incompatibilist: But even in the brain "with perfect free will" it is still physics that determines what goes on, not consciousness.

Compatibilist: I despair. What you fail to appreciate is that, in the brain with perfect free will, physical processes invariably occur in such a way as to enable the person to think her thoughts, reach her decisions, and do what she decides to do. The structure and working of the brain is such that physical processes going on it occur persistently in just the way required for the person to be in control of her consciousness, her decisions, her actions.

Let me, however, respond to what may lie behind your objection. In describing head processes in terms of their control aspect – as perceptions, desires, feelings, beliefs, intentions, decisions and so on – we are in effect referring to *those head processes and states, whatever their nature may be, that play the relevant role in controlling action*. Given physicalism, this means that references to head processes described in terms of their control aspects, are in fact references to those physical processes going on in the brain that are in fact playing the crucial role in determining what actions are performed (in the given environment). It is just that this reference is somewhat opaque, left open as a result of ignorance of the real nature our inner mental states. (By contrast, if Cartesian dualism were true, then descriptions of perceptions, decisions to act and so on would be references to and descriptions of non-physical, *mental* states and processes, not brain processes.)

Granted that physicalism is accepted, we can give just a little more theoretical precision to descriptions of our inner experiences. In referring to my decision to get up and greet my friend I can say I am referring to that *physical process* that occurred in my brain which played the crucial role in initiating and guiding my action, whatever precisely it might be.

Incompatibilist: This reply just might be acceptable given a functionalist account of consciousness. It does not work, however, given the view of experiential physicalism.

Functionalism is the view that mental aspects of brain processes are nothing more than control aspects – or *functional* aspects, as functionalists call control aspects. Given that a computer is playing chess, let us suppose, and is pondering its next move, what is going on inside the computer may be described in two different ways. On the one hand, there is the description of the *physical* processes occurring – electricity flowing through a complex network of transistors. On the other hand, there is the description of the *functional* or *control* aspects of these physical processes – that aspect that has to do with the task the computer has been set: to win the game of chess. Thus, in terms of this latter kind of description, the computer might be said to be "exploring the consequences of sacrificing a knight", or "thinking about moving the king's pawn up one square". According to functionalism, there is no essential difference between functional descriptions of computer processing, where there is no question of consciousness or sentience being involved, and functional descriptions of person's inner mental states and acts, where consciousness very definitely *is* involved. According to functionalism, when a control system becomes sufficiently sophisticated, sufficiently capable of representing its own inner functional states and processes, it becomes "conscious". The difference between a non-conscious and a conscious control system is one of degree, according to functionalism, not of kind. There are no extra, non-physical, experiential features of brain processes – or rather, if there are, they can, in principle, be fully understood in purely functionalist or control terms.

Granted functionalism, your account, *Compatibilist*, as to how the mental aspects of brain processes really can initiate and control action makes sense. For, according to functionalism, the mental aspects of brain processes are nothing more than functional or control aspects of those processes. Thus, in referring to mental aspects of decisions to act, we are, indeed, referring to the physical processes going on in the brain that initiate and control the relevant actions, but we are doing so in an opaque way, so that the decisions are not *specified* to be brain processes. But all this changes the moment functionalism is rejected, experiential physicalism is accepted, and it is accepted that (some) brain processes have an experiential aspect which cannot, even in principle, be reduced to, or explained in terms of, physical description. In referring to a decision to act as an inner mental act we refer to the experiential aspect of the brain process, and not to the physical aspect. And the experiential aspect is impotent to effect change. It is exclusively the *physical* aspects of our inner experiences that can play a role in determining what we do; the *experiential* aspects play no role whatsoever. Thus, granted experiential physicalism, we have no free will.

There is a paradox here. It might seem that experiential physicalism and the dual control hypothesis make free will possible, whereas functionalism does not, in that it reduces us to the status of robots. (We are, according to functionalism, no more than soft robots.) But actually, as we have seen, just the reverse is the case. Functionalism gives a real role to decisions in initiating and controlling action, whereas experiential physicalism does not. Being conscious in a way which cannot be fully explicated functionally, seems actually to undermine, rather than enhance, the possibility of free will.

Compatibilist: According to experiential physicalism, the mental or experiential aspect of a brain process *is* the control aspect of that process, and hence no special difficulty confronts the view that, in referring to conscious decisions to act, we are referring, somewhat opaquely, to physical processes involved in initiating and controlling our actions. A conscious decision to act *is* a brain process which, in turn, *is* a physical process that plays a decisive role in bringing about the bodily movements that constitute the action the person has decided to perform.[19]

Incompatibilist: This doesn't really meet my objection. Let me put it like this. Consider two purposive beings, Karl, a conscious human being, and Robot – a robot built and programmed to act just like Karl, but without a whiff of inner consciousness or sentience. Robot can do all the sorts of things that Karl can do, and furthermore, like Karl, can "think", "ponder", "imagine", "plan", "desire" "feel", have "hopes" and "fears", and "make decisions". The only difference is that, whereas Karl's thoughts, feelings, desires and decisions are (mostly) conscious, Robot's are not. But, according to the dual control hypothesis, this difference cannot

---

[19] See notes 16 and 17.

mean that Robot has less free will than Karl. All the "control" requirements for free will satisfied by Karl, are also satisfied by Robot. Even though entirely lacking in consciousness, Robot must, it seems, have just as much free will as Karl, according to the dual control view. This, on the face of it, seems like a *reductio ad absurdum* of that view. Consciousness must, after all, be an essential requirement for free will.

Compatibilist: Agreed: no free will without consciousness. The flaw in the above argument is the assumption that Robot is not conscious. According to the version of the brain process theory presupposed by the dual control hypothesis, what matters, for sentience and consciousness, is the *control* or *functional* features of a brain, not the substances out of which the brain is made. If the control features of Robot's brain are, in the relevant respects, similar to those of Karl, then this means, as a matter of fact, that Robot is conscious too.[20] The fact that Robot's brain is made of transistors or micro chips, while Karl's is made of neurons, does not deprive Robot of any consciousness of the kind that Karl possesses – as long as the control features of Robot's brain are similar to Karl's, and thus adequate to support consciousness, and conscious action.[21]

---

[20] It may be that, for a brain to become conscious, it is necessary that the brain *grows*, the growth of the brain being influenced by the way it functions. It may be, in other words, that we *grow* into full human consciousness. This may mean robots cannot in practice be conscious – unless a way is found for a being made of silicon microchips to grow.

[21] This is not the same as behaviourism – the doctrine that one is conscious if one behaves as if conscious (as we saw on pages 81-2). Suppose there is a robot that behaves as if conscious. Instead of a brain, however, the robot's motor and sensory impulses are transmitted to a nearby computer, which has, in it, a model of a possible conscious brain for the robot. The computer receives radio signals from the robot concerning sensory information; the computer then calculates how the robot's brain would have responded, if it existed, and then transmits the results to the robot, which then acts on the information received. As a result, the robot acts as if conscious. Nevertheless, it is not conscious because it does not have a brain, not even in the computer, where only a *model* of a brain exists. Calculations as to how a conscious brain would act if it existed are not the same as the workings of an actual conscious

Incompatibilist: There is still a difficulty. Given your view, consciousness seems to add nothing to freedom, and may even detract from it. In the case of a hypothetical non-conscious Robot, with a brain functionally similar to that of a conscious, freely acting human being, the Robot would seem, according to compatibilism, to be every bit as free as Karl, the human being. Indeed freer, because in the case of Robot, there is nothing in the brain other than physical processes guiding Robot's actions. His functional "desires", "intentions" and "decisions" really do play a crucial role in producing Robot's actions. In the case of Karl, it is not at all clear that his conscious, experienced desires, intentions and decisions play a similar role. These experiential processes cannot just be identified with functionally described neurological processes going on in Karl's brain. In principle, a second party, Jennifer, might know that such and such a complex neurological process occurring in Karl's brain is – functionally speaking – the decision to walk out of the room, but Jennifer cannot know what Karl knows in experiencing this decision, making the decision. What Karl experiences is thus extra to the functional aspect of the decision, the brain process, and thus not implicated directly in producing the action of walking out of the room. Whereas Robot's functional features of his brain processes are directly involved in producing his actions – as Karl's functional features of his brain processes are too – Karl's experiential features seem to play no role. There is no role for them to play. Compatibilist experiential functionalism is, in other words, in the same boat as epiphenomenalism: it can give no real role to conscious, experienced desires, intentions and decisions in producing actions. This role is completely filled by brain processes *functionally construed and described* – according to the compatibilist view under consideration.

Compatibilist: But what would it be for us to be more directly aware of the control or functional aspect of our brain processes than we ordinarily are? It would be no good to be directly aware of the *neurological* aspect. In that case, as I have already remarked, we would become hopelessly lost among the neurons

---

brain, even if the behaviour that results is the same.

and synaptic junctions of our brain. We would not know what we need to know: the *control* aspect of our head processes. Would we be more directly aware of the control aspect if we became directly aware of the content of perception without any sensory experience whatsoever, so that we posses consciousness but no sentience?

The phenomenon of blindsight might indicate what this would be like. Some blind people are able to guess correctly about the nature of objects placed before them, even though they see nothing, and have no visual experience whatsoever. These people are blind because the primary visual cortex, at the back of the head, has been damaged. Information from the eyes is able to reach the brain via the optic nerves by another route, and it is this which enables the person to know, without perceptual experience, what has been placed before him.

Would we be more directly aware of the control aspect of our brain processes if all our perceptual knowledge was analogous to blindsight? We know directly what is before us, what is said, how we are related to our environment, but see nothing, hear nothing, feel nothing. Such a mode of existence would surely be an absolute nightmare. One would have knowledge of one's surroundings, knowledge of one's body in one's surroundings, but one would, it seems, be bereft of any idea as to how one has acquired this knowledge. Initiate certain impulses – corresponding, for example, to turning the head – and, abruptly and mysteriously, one's knowledge of one's surroundings is transformed.[22] It may well be that consciousness devoid of sentience is physiologically and psychologically impossible.[23] But if it is possible, it would not seem to involve a more direct knowledge of the control aspects of brain processes than we ordinarily have.

The attempt to devise a kind of knowledge of control aspects of our brain processes more direct and accurate than our ordinary knowledge seems to fail. The outcome of the attempt is, indeed, to

---

[22] I wonder whether such a horrifying existence could be made the subject of a short story.

[23] It might even be logically impossible (in that the very idea of consciousness presupposes some sentience).

impress on one just how direct and accurate our ordinary knowledge of the control aspects of our brain processes is. It seems impossible to devise anything more direct and accurate. And this is, perhaps, not surprising. Natural selection might be expected to design brains good at acquiring this kind of self-knowledge, vital for survival. An animal does not need to know anything about the neurological workings of its brain, but it does need to know what this perception of danger, or of food, betokens.

Much depends, in this context, on whether one adopts the "internalist" or "externalist" account of perception, mentioned above and discussed in chapter three. The internalist account depicts perceptual sensations as deceptive about the nature of external reality and, at the same time, utterly distinct from associated brain processes. This is all but indistinguishable from epiphenomenalism. The externalist account, by contrast, depicts perceptual experiences (which are not illusions or hallucinations) as accurately representing external reality. Grass really is green, and honeysuckle really is yellow and sweet smelling. Furthermore, what we know about the process occurring in our head is precisely the control or functional aspect of that process: what we need to know in order to know how to act in response to what we perceive. In experiencing the visual sensation of yellow honeysuckle when blindfold, induced perhaps by a probe to the visual part of my brain, all I know about what is going on inside me is: "this is the sort of process that would occurs inside me if I really was seeing yellow honeysuckle". Externalism holds, in short, that our knowledge of our inner perceptual sensations is based on our perceptual knowledge of things external to us – what we require in order to know how to act, the *control aspect* in other words.

Incompatibilist: I remain entirely unconvinced. Let me try again. If it is logically possible to have Robot acting like a conscious, freely acting person, with a brain functionally similar to such a person, but Robot is nevertheless devoid of consciousness, then it would seem purely functional descriptions of Robot's brain states would suffice to explain his actions. If this is the case, adding on consciousness can make no addition to free will whatsoever. The conscious aspect of brain processes can play no role in producing actions in addition to that of brain processes conceived of functionally, with conscious aspects ignored. Experiential physicalism does seem, in this respect, indistinguishable from epiphenomenalism.

In order to avoid this conclusion, it seems obligatory to hold that Robot, as just conceived, is not possible, even in principle. If Robot acts as a conscious, free person, and has a brain functionally similar to the brain of such a person, then Robot is conscious and free. Robot must, in other words, in these circumstances, be conscious and free – at least, as conscious and free as the corresponding person. But this means that, from a purely functional, non-experiential description of Robot, of his actions and brain processes, an experience-laden description can be deduced. It is not logically possible to have the one without the other. And that would seem to go against the whole idea of experiential physicalism, which holds that the experiential cannot be deduced from the non-experiential.[24]

Is Robot a logical possibility? Then there seems to be no role for consciousness in free will. Is Robot not a logical possibility? Then, from a purely functional description of Robot it must be possible to deduce that Robot is conscious. But this goes against a basic premise of experiential physicalism. Either way, experiential physicalism fails to explain how consciousness can contribute to free will.

Your only reply, *Compatibilist*, to this criticism is to argue that Robot is a *logical* possibility, but not a possibility *in fact, in practice*. Robot may exist in a logically possible world, but cannot exist, as a matter of fact, in our world. But this reply is inadequate, in at least two respects. First, the mere logical possibility of Robot is enough to establish that consciousness plays no active role in free will. It may be that, in our world, whenever a being acts as if conscious, and has a brain like that of a conscious person, then that being is, as a matter of fact, conscious. Nevertheless, consciousness is not logically necessary for free action, and that

---

[24] In chapter 3 indeed we had something like a proof of this proposition, with the argument that physics cannot predict that ripe tomatoes (for example) are (experientially) red.

234

suffices to show experiential physicalism cannot assign an active role to consciousness in producing free action. Secondly, if beings like Robot are invariably conscious as a matter of fact, in our world, then there must be some reason for this, some explanation. It cannot be a mere accident. Brain function must be linked to consciousness in a way that is somewhat analogous to the way releasing stones near the earth's surface is linked to the stones falling to the ground. In this latter case, there is an explanation: Newton's law of gravitation. So, in the case of Robot-like beings always in fact being conscious, there must be some kind of explanation. But what explanation could there be? What theory of consciousness could predict and explain that a brain operating in such and such a way is invariably conscious?

Compatibilist: You have almost convinced me that you are right, and my dual-control, experiential physicalist account of free will is untenable. Almost, but not quite. Let me try to reply to your criticisms.

To begin with, beings like Robot, acting as if conscious, and with brains functionally similar to those of conscious persons, but not conscious, are called *zombies* in the philosophical literature – so let's adopt that terminology here.

A preliminary point to note is that it would not be logically possible for a zombie to do much of what we conscious human beings do. Much that we do presupposes consciousness. I cannot share a joke with you, praise or scold you, or even ponder a question, if I am a zombie. But this is a rather trivial, verbal point. It suffices for the incompatibilist criticism that the zombie can imitate these actions.[25] We do not require that the zombie can actually *do* them.

Is there a knock down argument that establishes that zombies must be logically possible? The point that physics cannot predict experiential features of things, like redness, does not suffice. For what has to be demonstrated is that a purely *functional* description of a potential zombie – his actions and brain processes – cannot imply consciousness. A *functional* description is not the same as a *physical* description. And implying that this being is *conscious* is

---

[25] In a sense of "imitate" which does not itself presuppose consciousness.

not at all the same as implying that the being has these *specific* experiences (the visual sensation of redness, etc.) Delicate questions arise, too, in connection with how the functional description is to be formulated. Does the functional description include experiential terms, or not? If we assume that it does not, then we must conclude, in my view, that no purely functional description of the potential zombie can imply that he is conscious. The zombie must be a logical possibility.

The only remaining hope, it seems, is to hold that the zombie, though *logically* possible, is not *in fact* possible. Consciousness being what it is, in this world, if you have a brain that functions like that of a conscious person's brain then, as a matter of fact, you are conscious. And it is this fact about consciousness in our world that ensures that consciousness plays a vital active role in free action. All those astonishing free actions we perform, all those things we freely do and create, demand consciousness. Zombies cannot, *as a matter of fact*, in this world, do them, or "imitate" doing them, even though it is *logically possible* to have zombies that do such things.

Incompatibilist: But what about my point that there must be some *explanation* as to why a brain that functions as if conscious, must in fact be conscious?

Compatibilist: It is tempting to shelter behind our ignorance about how the brain works, how it enables us to be conscious. When we know and understand more neuroscience, we may think, all will become clear. There are many examples in the history of science of agonizing philosophical or conceptual problems dissolving with the advance of scientific knowledge.[26] But this is not very convincing. It is not clear how advances in the neurosciences could conceivably come up with the kind of explanatory theory we require.

Let us suppose there is a theory, H, which explains why any potential zombie, Z, is sentient and conscious. H, together with a

---

[26] Here is an example. At the time of Newton and Leibniz, there was a major dispute as to whether "quantity of motion" should be defined as mv, or $mv^2/2$ (where m is the mass of a body, and v is its velocity). Gradually it was realized that both are important, the first being momentum, the second kinetic energy.

neurological and purely functional description of the brain processes and actions of Z, logically imply that Z is sentient and conscious. Two questions immediately arise about H.

First, it is difficult to see how H can be tested. We have a potential zombie, Z, before us. H predicts that Z is conscious. If Z is conscious, he knows he is. But how can he communicate this information to the rest of us? Whether he is or not makes no difference. Either way, he will tell us he is conscious. The rest of us have no way of checking up on the correctness, or falsity, of this prediction of H. In a sense, indeed, we have no way of bringing into the public domain knowledge of how consciousness is distributed among humanity. I cannot refute the hypothesis that I alone am conscious, and the rest of you are zombies – anymore than you can refute the hypothesis that you alone are conscious.

Quite apart from this epistemological problem facing H, there is the argument of chapter three of this book (which I, a character in this book mysteriously seem to know all about) which seems to show that H, if it exists, must be so horrendously complex that it would not be explanatory. An explanatory H, in other words, is impossible. And indeed no one has, it seems, been able to think up even a possible H that might explain why brain processes and conscious inner experiences are correlated in the way that they are.

As it happens, our author has put forward just such an explanatory H.[27] The idea can be put like this. Our various kinds of sensation – visual, auditory, olfactory, tactile – seem entirely different, one from the other, entirely arbitrary. It seems impossible to guess, from having experienced one sensory mode, the nature of others. Consider a person who hears a specific kind of sound, played on a violin say, of specific timbre and loudness. Suppose now that the pitch of the note is continuously increased. It would be easy for the person correctly to conceive what note comes next. But nothing comparable is possible if one has to conceive, from auditory sensations, what visual sensations are like, never having had experience of them before. Indeed, even within one sensory mode, visual say, or auditory, it would be all but impossible to conceive, on the basis of limited experience, what

---

[27] See my (2001), pp. 126-129; and my (2006b).

hitherto unexperienced sensations are like. If one has only seen red, orange and yellow, what basis does one have for imagining and anticipating blue?

But let us now think of all possible sensations – sensations experienced not just by all human beings, and not just by all sentient animals, but sensations experienced by all sentient beings that are *in fact* possible, whether they will ever exist or not. It could be that this vast universe of possible sensations can be ordered into a multi-dimensional space in such a way that, as one goes through the space, sensations vary smoothly and predictably, like the violin note increasing in pitch. It might be that all possible sensations can only be ordered smoothly in this fashion in one possible way. The sensations we experience come from widely separated patches in this vast, uniform space of possible sensations: that is why they seem, relative to each other, so arbitrary and unpredictable.

Consider now all possible brain processes and structures corresponding to these sensations, one functionally described brain process corresponding to each sensation. Let us suppose these can be arranged in a space of all possible, functionally described, sentient brain processes in such a way that brain processes vary smoothly as one moves through the space – it being possible to do this in only one way. Now suppose that these two spaces are matched up (i.e. put into one-to-one correspondence) in such a way that if two brain processes are close together in the one space, then the corresponding sensations are close together in the other space. Let us suppose that this "distance preserving" mapping can be done in only one way. The spaces cannot be rotated with respect to each other, and brain processes in the one space cannot be moved with respect to the corresponding sensations in the other space, without the "distance preserving" mapping being violated. The "distance preserving" mapping is, in short, unique.

And now suppose that this unique mapping, between all possible sentient brain processes and all possible sensations, is precisely the one which allocates, to each sensation, the brain process that *is* the sensation. If the brain process occurs in an appropriate brain, then the person experiences the corresponding sensation. At once we have an explanation as to why brain processes and sensations are correlated in the specific, fixed way that (we are presuming) they are. If brain processes and sensations were to be correlated differently, the unique "distance preserving" mapping between the two spaces would be violated. Brain process-sensation correlations are fixed rigidly.

A "unique matching" hypothesis, H, along these lines could conceivably be true.[28] If true, it would explain why brain processes and sensations are correlated in the way that they are. Even if H is false, it is still significant, because it indicates it is possible, after all, to have a theory that explains brain process-sensation correlations, and thus predicts that potential zombies are sentient and conscious.

One may doubt, however, that any such H is really necessary. The moment we put the zombie problem into the context of the broader problem of why brain processes and sensations are correlated in the fixed way they seem to be, doubts begin to arise as to whether it even makes sense to think of such correlations changing. For, suppose they do change, in any way one pleases. Everything physical remains the same, but for some people, or for everyone, the spectrum of colours is inverted, let us say (so that what was red is now blue, and *vice versa*), or what was visual is now auditory and *vice versa*, or pain and pleasure are interchanged. Nevertheless, granted experiential physicalism and the persistence of everything physical, everything would proceed exactly as before. No one would say anything about these remarkable changes, or react in such a way as to indicate that they had even noticed them. People would smile with pleasure when in agony, and howl with pain when experiencing delight. At no point would anyone say or do anything to indicate that the experiential world had been turned upside down.

If this really did occur, the utter incapacity of experience to influence action would become all too apparent. It would be very clear indeed that we are wholly in the grip of physics, experience no more than a sideshow, entirely impotent. This is the epiphenomenalist nightmare.

---

[28] It would be very difficult to test.

The problem posed by the logical possibility of the zombie is, it seems, just a special case of the much more general problem posed by the logical possibility that the relationship between the physical and the experiential changes *and yet, in a certain sense, everything remains the same.* No one can say anything about the changes, or respond to the changes, in any way whatsoever (because everything physical remains the same), even though the changes are as radical as one cares to imagine. The special case of the zombie restricts these changes to the zombie: there are no inner experiences associated with his brain processes.

If such changes in the customary relationships between the physical and the experiential are logically possible then, so it would seem, our inner experiences can play no role in influencing action. For, in that case, we would do exactly what we in fact do *even if our inner experiences were entirely different.* Some way must be found to declare that the nightmare of arbitrary changes in the relationship between the physical and the experiential (everything physical remaining the same) to be *impossible.* If it is impossible, then the relationship between the physical and the experiential could not be other than they are, and the argument for the impotence of the experiential breaks down.

There seem to be just two ways of arranging for this impossibility.[29] The first holds that such arbitrary changes in correlations between the physical and experiential are *logically* possible, but not possible *in fact.* They are possible in some imaginary, logically possible world, but they are not possible in fact in *our* world, *this* world. The second option is more radical. It declares such arbitrary changes to be not even *logically* possible. Here, now, are two views of the first kind, and then two views of the second kind.

1. Accept experiential physicalism, the idea that inner experiences are contingently identical to brain processes, *and* the view that there is a hypothesis H – like the one indicated above – which *explains* why physics and the experiential are correlated in the way that they are. Given the truth of the explanatory

---

[29] Throughout we are assuming physicalism, and thus ignoring such options as that a distinct mind interacts with the physical brain.

hypothesis H, it can be argued that, even though it is *logically possible* that physics and the experiential might be correlated differently from the way they are, this is not *in fact* possible. It is denied by H, which explains why the physical and the experiential are correlated in the way they are. In a *logically possible* world, physics and the experiential might be differently correlated, but in *this world*, this is not in fact possible. This in turn means it is not in fact possible for physics and the experiential to be correlated differently: the argument for the impotence of the experiential collapses. This is especially so if inner experiences and decisions to act are contingently identical to brain processes – these processes nevertheless having experiential or mental features which we can only become aware of by having these processes occur in our own brain.

2. Accept experiential physicalism but *reject* H. That is, deny that any such H exists. Instead, hold that the true physical theory of everything, T, can be supplemented with a vast number of additional postulates (perhaps infinitely many) linking the physical and the experiential to form a true theory T*. T* is too complex to be explanatory. However, T*, like T, can correctly make hypothetical or counterfactual predictions of the form: a person with such and such a brain in which such and such a process occurs would experience such and such a sensation. The truth of T* suffices to ensure that arbitrary changes in physical/experiential correlations are not *in fact* possible, in our world. This in turn ensures that our conscious decisions to act really can play a substantial role in bringing about our actions.

3. Accept functionalism – the doctrine that inner experiences and states of awareness are nothing more than certain sorts of processes occurring in certain sorts of brains *functionally specified* (i.e. specified in control terms). If functionalism is true, then keeping the physics intact but changing the way the physics is related to the experiential becomes impossible because any such change would require a change in brain processes *functionally described*, which would require a change in the physics of these processes. Put another way, functionalism denies that there is any such thing as inner experience over and above brain processes, so of course a

change in the relationship between the experiential and the physical is logically impossible.

4. Accept experiential functionalism, but hold that the experiential aspect of a brain process is nothing more than what we learn when we have the process occur for the first time in our own brain. Thus the visual sensation of blue just is a specific kind of brain process, specified in functional or control terms, X let us say. A person blind from birth who has her sight restored learns something new when she first experiences the visual sensation of blue, namely "what it is to have X occur in your own brain". Of course she doesn't know that this is what she knows. (What we would ordinarily know about what is going on inside us when we experience blueness is that this is the sort of thing that happens to us when we see blue things – the sky in the day, for example, when there are no clouds.) Nevertheless, even though she doesn't know that she knows this, it remains true that this is what she knows: what it is to have X occur in her brain. But if this is what the visual experience of blueness is – having X occur in your brain – then there is no possibility of the experience of blueness being associated with a different kind of brain process, Y say (the visual sensation of redness). The blue visual experience is not like a layer of blue paint on X, which might be stripped off X and applied to Y instead. "What it is to have X occur in your own brain" can only be learnt by having X occur in your own brain. It cannot be learnt from a description of the physical or functional features of X, or from an examination of X as an external process (in someone else's brain), however exhaustive. X must occur in one's own brain. Only then does one know what it is to have X in one's brain – that extraordinary blue visual sensation one associates with sky, sea, bluebells and forget-me-nots. But this blue sensation could not be associated with a different process, Y, because it is just "what it is to have X occur in your own brain".

This view is not functionalism (3 above): it holds that one really does learn new things in having certain kinds of processes occur for the first time in one's own brain. Inner experiences, inner states of awareness, *are* brain processes, but what we learn from having these processes occur in our own brains cannot be derived from physical or functional descriptions of these processes, however complete. On the other hand, this view does not need to appeal to the explanatory hypothesis H (1 above) or the non-explanatory theory T* (2 above): what we learn when we have X occur in our brain for the first time is what the sensation of blueness is, but this sensation cannot be associated with a different process, Y say, because it is just "what it is to have X occur in your own brain". To assert that the sensation of blueness might be correlated with Y would be equivalent to saying "having X occur in your brain might be having Y occur in your brain", a logical contradiction. Thus it is not possible for the sensation of blueness to be anything other than X – and no explanatory H is required to establish this impossibility. H is not required.

Option 3 deserves to be rejected. It is equivalent to denying the reality of inner experience, the experiential world. Option 1 is dubious, because H is dubious. It is not just that H is wholly hypothetical. It remains doubtful that any such H can be given a coherent formulation. Option 2 is perhaps the best bet, in that places the least strain on our credulity. But option 4 is an intriguing possibility.

Incompatibilist: Option 4 is pure sophistry. No logical contradiction is involved in the assertion "X may be occurring in my brain, but it is, for me, just as if Y is occurring" – where Y, let us assume, is ordinarily the visual sensation of redness. Granted this fourth view, in other words, it is entirely possible logically for correlations between the physical and experiential to be changed arbitrarily.

Compatibilist: But is it? If experiencing the visual sensation of blueness just is having X occur in your brain, and nothing more than that, how could it possibly be having Y occur in your brain? That would be equivalent to saying what it is to have X occur is what it is to have Y occur, when these are quite distinct, and distinguishable.

Incompatibilist: Sophistry! My objection is very simple. Given experiential physicalism, the visual sensation of blueness is either necessarily identical to X, or contingently identical to X. If necessarily identical, then it is not logically possible for the sensation of blueness to be anything other than X, which must be wrong since it could be some brain process other than X. It is not a

243

logical contradiction to say the sensation of blueness is Y. If, on the other hand, the sensation of blueness is contingently identical to X, then it follows, immediately, that this sensation might be identical to something other than X. Option 4 is untenable.

Compatibilist: You may be right. But could one not say that it all depends on what one takes "the visual sensation of blueness" to be? If it is taken to be "what one knows about what is going on inside one's head whenever one sees something blue" the identity is contingent. The visual sensation of blueness, so construed, is X but might be something else. If, on the other hand, it is taken to be "what one knows about what is going on inside one's head whenever X occurs in one's head" the identity is necessary. The sensation, construed in this way, is necessarily X.

Incompatibilist: Sophistry! Sophistry! I can't bear it. My head is hurting. Stop!

I want to turn to another difficulty. Granted experiential physicalism, everything that occurs can in principle be predicted and explained physically.[30] This includes everything to do with human life. Thus the real explanation as to why we do what we do has nothing to do with *us*, with our intentions, desires or decisions. It is just physics. And that means we are the playthings of physics, free will nothing but an illusion.

Compatibilist: A full physical description and explanation of a person's actions – supposing that such a thing is possible – would include a precise specification of that person's evolving intentions, desires and decisions to act. These would be described as physical processes, not as the inner experiences, intentions or decisions of the person in question. Nevertheless, these physical processes, occurring in the brain of the person in question, would be his thoughts, intentions, decisions.

Incompatibilist: But, for free will, this is not enough. What is required is that the person – Gordon Brown let us suppose – can be understood as a person, acting freely, in charge of what he does, responsible for his actions and for his decision-making. Any such explanation or understanding of Brown as a free agent becomes

---

[30] Not of course in practice, as we saw in chapter three, note 2.

redundant if physics in principle explains everything that goes on. Physics cancels free will.

Compatibilist: But this just assumes that if physics in principle provides an explanation for everything that occurs, there cannot also be a true *personalistic* explanation – as we may call it – which explains what goes on in terms of the desires, beliefs, intentions, decisions of the person in question – and the human context in which he acts. But this assumption is false! Corresponding to the *dual control hypothesis* there is a *dual explanation hypothesis*. Human beings are such that they are amenable to being explained and understood simultaneously in two very different ways: physically and personalistically. People exhibit what may be called *double comprehensibility*.[31] A person can be understood correctly as acting freely even though, in principle, everything that occurs in connection with that person's actions could in principle simultaneously be explained and understood physically. The latter does not cancel the possibility of the former.

Double Comprehensibility

At this point I, the author, need to take over from our two disputants. Double comprehensibility needs a bit of background exposition to be understood properly. This is something that I, the author, ought to try to provide. In what follows, we shall see that reformulating the dual control hypothesis as the dual explanation hypothesis – or the thesis of double comprehensibility – strengthens the argument for compatibilism in two important ways.

A key feature of explanation, quite generally, is that it tells us, not just what actually occurs, but also what would occur if conditions had been different in a variety of specified ways. Physical explanation achieves this by being compounded of two distinct parts: (a) a theory, and (b) initial conditions. These, taken together, imply subsequent states of affairs. The theory predicts that one instantaneous state of affairs – the instantaneous state of the solar system perhaps – is followed, some time later, by a different instantaneous state of affairs. This prediction is

---

[31] See Maxwell (2001), ch. 5.

explanatory because the same theory predicts that if the initial conditions had been different, in a wide range of ways, then the outcomes would have been different, in a wide range of ways. Newtonian theory does not just tell us how stones that are actually thrown travel through the air. It tells us how stones would travel were they to be thrown with any possible initial velocity, even though they have in fact not been thrown at all. It is this which makes Newtonian theory explanatory – together with the *unity* of the theory (the *same* laws applying to the wide diversity of phenomena to which the theory applies).

Something analogous holds for what may be called *purposive* and *personalistic* explanations. As we have already seen, anything that pursues goals, whether person, animal, robot or guided missile, must be able to respond in a variety of ways, to a variety of circumstances. Only then can the thing in question achieve its goal in a variety of conditions. This means that the actions of any goal-pursuing thing are comprehensible and explainable in a distinctively purposive kind of way. A factually correct specification of the nature of the purposive thing in question – its goals and capacities – plus a specification of what occurs in its environment, will imply what the thing does in pursuit of its goal. Even if the purposive thing fails to achieve its goal, nevertheless a prediction may still be forthcoming concerning what the thing *attempts* to do in pursuit of its goal. In addition – and this is what makes it an *explanation* and not just a *description* – the specification of the nature of the goal-pursuing thing carries implications about a variety of things this thing *would* have done in a variety of *different* circumstances. If the purposive thing in question happens to be Gordon Brown trying to decide whether or not to call an election, a knowledge of Gordon Brown's psychological make-up, goals and priorities makes it possible to predict what Gordon Brown would do in a variety of different circumstances. (If he had been ahead in the polls by a big margin, then he would have called an election; if the polls made it clear that he would lose, then he would not call an election; etc.)

We need to distinguish *personalistic* and *purposive* explanations. A personalistic explanation is the kind we ordinarily employ in understanding others – and ourselves. It involves imagining you are

246

the other person, with that person's desires, beliefs, feelings, values, goals, situation, problems, plans, activities, relationships, and so on. In seeking to explain and understand another personalistically, you imagine you are that other person, with that person's problems, aims, temperament, and so on, and you then work out what you would do, how you would feel, what you would want, as that other person, just as you would do that for yourself, but taking into account all the differences between yourself and the other person.[32]

Understanding others personalistically, and understanding yourself – knowing yourself, your desires, plans, hopes, fears, etc. – are, on this view, in a sense, two sides of the same coin. We become self-conscious persons, aware of what we desire, feel, believe, etc., in part because we have interacted with others from babyhood onwards, becoming aware of the intentions and emotions of others. We are capable of understanding ourselves – we have a self-conscious self to be understood – in part because, from an early age, we have understood others; and in understanding others we do so as we understand, or know, ourselves.[33]

Personalistic explanation is often called "empathic understanding" and, among psychologists and philosophers, goes by the name of "folk psychology" and "theory of mind" – although what is usually meant by these terms differs in important respects from what I take personalistic explanation to be, as we shall see shortly.

A purposive explanation applies to any goal-pursuing entity, whether sentient, conscious, or not. It explains what the entity

---

[32] For more detailed accounts of personalistic understanding see Maxwell (1984, pp. 174-81, 183-9, 264-75; 2001, pp. 103-112, 188-9.)

[33] Elsewhere (2001), pp. 188-9, I have argued that we acquire self-consciousness as a result of becoming aware of others' personalistic understanding of ourselves. Seeing ourselves from outside, in this way, compels us to take note of what is left out – our own inner experiences and consciousness. Thus consciousness of our own consciousness is born. But this view faces a problem. Those who are autistic lack the capacity to acquire personalistic understanding of others, and so gain knowledge of what others think of them. Does this mean that severely autistic people are not self-conscious? Or should one rather hold that the self-consciousness of the autistic refutes the above account of the genesis of self-consciousness?

does by characterizing the actions of the entity as being designed to achieve the entity's goals in the given environment. But it does this without any hint that the entity in question is sentient or conscious – even if it is. Purposive explanations treat human beings as if they are zombies. If a purposive explanation employs terms that ostensibly presuppose sentience or consciousness, quotation marks must be used to indicate that neither are being assumed. Thus a purposive explanation may assert that an entity is 'trying' to achieve such and such a goal, 'believes', 'knows', 'desires', 'values', or 'fears' such and such an eventuality, or is 'struggling to solve' such and such a problem. These terms are to be understood in just the way they would be understood if applied to a thermostat or guided missile, where there is no question of sentience or consciousness whatsoever.

The big distinction between a personalistic and purposive explanation, then, is that a personalistic explanation tells you what it would be for you to be the thing explained, whereas a purposive explanation does nothing of the kind. A personalistic explanation relates the thing explained to yourself, and is, to that extent, inherently anthropomorphic. All personalistic explanations are also purposive explanations, whereas no purposive explanation is personalistic.

We can now indicate what it is to act freely like this: A person acts freely if there is a true, freedom ascribing personalistic explanation of the person's inner and outer actions. This in turn requires, at least, that the kind of conditions, already discussed, are satisfied. The person's authentic self must be in control, decisions to act must actively guide or control actions decided upon, and these decisions must be the person's own, not foisted onto him by external agents (by means of brainwashing, manipulation, or chemical or neurological disorders).[34]

So far everything is even handed between *compatibilism* and *incompatibilism*. But at this point the two views clash.

---

[34] It is important that the true, freedom ascribing personalistic explanation specifies the *real* reasons and motives for the action in question, and does not merely specify declared or official reasons or motives (which may be hypocritical, mere rationalization). This is part of the requirement for the personalistic explanation to be *true*.

Incompatibilism demands, for free will that, in addition to the above, it must *not* be the case that everything the person thinks and does could, in principle, be predicted and explained *physically* – even if only probabilistically, and even if only when described in purely physical terms. *Compatibilism*, by contrast, holds that free will is entirely compatible with everything being capable, in principle, of being predicted and explained in purely physical terms.[35]

What is to be gained by reformulating the dual control hypothesis as the dual explanation hypothesis? There are two. (1) We can demand of the dual explanation thesis that, for free will, the true, freedom-ascribing personalistic explanation of the person's actions is not reducible, even in principle, to the physical explanation of these actions. (2) We can demand that, for free will, personalistic explanations are intellectually authentic and fundamental. These are two quite challenging demands that we may require compatibilism to satisfy that cannot even be formulated in terms of the dual control hypothesis. Let us consider them in turn.

(1) A basic *incompatibilist* objection to the dual explanation thesis is likely to be that, if physics can, in principle, explain everything a person thinks and does, then the existence of a true personalistic explanation of what he thinks and does can hardly provide grounds for holding the person acts freely. For everything explained personalistically can, it seems, be explained in a far more powerful and universal way by physics. Any personalistic explanation would apply only to a highly restricted range of circumstances that include both the continuing existence of the person, and the environment in which the person acts. The true fundamental physical explanation would include all this plus a vastly wider range of circumstances as well. In addition, whereas any personalistic explanation could hardly predict what happens with certainty (since something from beyond the given environment – a meteor, for example – might unexpectedly crash

---

[35] I defend compatibilism appealing to the idea of dual explanation in my (2001), ch. 6. This has been criticized by David Hodgson: see McHenry (2009), pp. 199-216. For my reply, see McHenry (2009), pp. 259-65.

in and kill the person), the true physical explanation *would* predict and explain, without exception. Personalistic explanations, being much more restricted and weaker than physical explanations, seem redundant.

This objection fails, however, if personalistic explanations are not reducible, even in principle, to physical explanations. For then personalistic explanations would not be redundant. Compatibilism is only valid, let us demand, if this is the case. But it clearly *is* the case. Personalistic explanations are massively and inherently experiential in character. In explaining the actions of another person experientially, we relate that person's thoughts, feelings, sensations, motives to our own. We understand the other as if we are ourselves the other, experiencing what the other experiences. But, as we saw in chapter 3, the experiential cannot, even in principle, be reduced to the physical. Hence, personalistic explanations cannot be reduced to physical explanations; they are not redundant. They tell us what even a complete physical explanation could not tell us, even in principle. The additional requirement for compatibilism to be valid is met.

A case can be made out for holding that even some purposive explanations are not reducible to physical explanations. A true purposive explanation of the actions of a purposive being might predict what the being would do in possible circumstances which are not *physically* possible. Thus, a purposive explanation of a devout person might predict what the person would do if he "saw" an angel descend from the sky in a manner which contradicts the laws of physics. Or a purposive explanation of the actions of a missile with an on-board computer guiding its actions might predict what the missile would do in response to another missile attacking it whose actions violate the laws of physics (because it travels faster than light, for example). In both cases, such counterfactual predictions would not, it may be argued, come within the province of the physical explanation, because the states of affairs in question contradict physics.

If purposive explanations are not reducible to physics, then this strengthens the case for holding that personalistic explanations are not so reducible (although the latter case holds even if the former one does not).

I conclude, in any case, that the additional requirement for compatibilism to be valid holds.

(2) *Incompatibilists* may well argue that personalistic explanations, even if not reducible to the physical, are nevertheless not intellectually authentic and fundamental in the way that physical explanations, couched in terms of fundamental physical theory, are. As a consequence, defending compatibilism by appealing to personalistic explanation fails.

Let us accept this argument, and hold that compatibilism is only valid if personalistic explanations can be shown to be intellectually authentic and fundamental – in some legitimate sense of "authentic" and "fundamental". We have here an additional demand that compatibilism must satisfy to be valid. I now argue that this demand cannot be met if knowledge-inquiry is accepted, but *is* met once wisdom-inquiry is adopted.

Granted knowledge-inquiry, personalistic explanations seem intellectually inauthentic, and certainly not fundamental. A genuine scientific explanation owes its authenticity to being (a) objective (b) impersonal (c) factual (d) rational (e) predictive (f) testable (g) empirically corroborated and (h) scientific, in that there is an objective, impersonal, factual scientific theory which predicts the phenomena to be explained, and is empirically testable and corroborated, and so scientifically and rationally appraised. But a personalistic explanation, amounting to one person understanding another by imagining she is that other, lacks all these features. It is (a) subjective (b) personal (c) emotional and evaluative and so not factual (d) intuitive, and thus non-rational (e) untestable (f) not corroborated empirically and (g) not scientific. No wonder psychologists and cognitive scientists, presupposing knowledge-inquiry, call personalistic understanding "folk psychology", and seek to replace it with authentic scientific explanation.[36]

Granted wisdom-inquiry, however, the intellectual status and authenticity of personalistic explanation is very different. For, according to wisdom-inquiry, articulating problems of living and

---

[36] Thus Paul Churchland declares "The term 'folk psychology' is ... intended to portray a parallel with what might be called 'folk physics', 'folk chemistry', 'folk biology', and so forth". See Churchland (1994).

proposing and critically assessing possible solutions (possible actions) is intellectually fundamental – in a sense more fundamental than the search for scientific knowledge and explanation. But it is just this intellectually fundamental activity that we need to engage in, in order to develop personalistic understanding of others. If I am to have personalistic understanding of you I need to know (1) what your problems of living are, (2) what can be done to solve them (3) what you take your problems of living to be, (4) what you will do in an attempt to solve them, and (5) what your capacities are to perform the relevant actions. All this is intellectually fundamental within the framework of wisdom-inquiry. Personalistic understanding is, according to wisdom-inquiry, (a) objective (b) personal (c) emotional and evaluative, but also factual (d) both intuitive and rational (e) fallibly predictive (f) capable of being assessed rationally, perhaps even empirically, and (g) not scientific but nevertheless presupposed by science.

The transition from knowledge-inquiry to wisdom-inquiry, in short, transforms the intellectual status of personalistic understanding: it ceases to be mere folk psychology, waiting to be replaced by proper scientific explanation, and becomes intellectually central, authentic and fundamental – or at least not eliminatable, even in principle.[37] Advances in psychology, cognitive science and neuroscience will undoubtedly enable us to improve our personalistic understanding of each other and ourselves, but will not replace it.

Scientific explanation of natural phenomena enable us to *manipulate* nature, mainly by providing the means to create technology. Personalistic understanding can be used to manipulate too, but its proper, primary use is to facilitate *cooperation*. If two or more people are to engage in a common endeavour cooperatively they must be able to understand each other's ideas,

---

[37] Within wisdom-inquiry, personalistic explanation is fundamental but not *more* fundamental than physical explanation. Each presupposes the other. Any personalistic explanation must make factual presuppositions about the environment in which the person acts. And, as I go on to explain in the text, any scientific explanation rests upon, presupposes, and is the product of, personalistic understanding between scientists.

aims, desires, objections, beliefs, problems, which means they must be able to acquire personalistic understanding of each other, as required for the matter in hand. This point is quite general; it applies to any cooperative endeavour, including the cooperative endeavour of science.

There are at least two very different reasons why I may seek to acquire personalistic understanding of you. First, I may be interested in improving my knowledge and understanding of *you*. But secondly, I may be interested in your beliefs about aspects of the world, not because I am interested in you, but because I am interested in *the world*, and I think your beliefs may contribute to or correct mine. Scientific knowledge emerges in this way, the product of scientists acquiring personalistic understanding of each other with the emphasis on *ideas about the world*, interest in the personal being suppressed, diverse theories and hypotheses becoming common currency in this way, and so open to cooperative criticism, empirical testing and improvement. The mere act of understanding a new theory put forward by another scientist, or the report of a new experiment or observation, implicitly involves and presupposes personalistic understanding, whether this is recognized or not. If natural science were pursued within the framework of wisdom-inquiry it would be obvious that it presupposed personalistic understanding in this way, its cooperative aspects being inconceivable without it. In fact science has long been pursued presupposing standard empiricism and knowledge-inquiry; this has rendered the personalistic dimension of science invisible, and has even led to the idea that personalistic understanding is no more than folk psychology, to be replaced by good scientific explanation.

This concludes my case for holding that free will, in a worthwhile sense, is possible within an experientially physicalistic universe. Our experiential beings are in part physical beings, an integral part of the physical universe, conscious decisions to act being neurological processes that do indeed control and guide our actions. Our apparently frail conscious decisions to act have all the potency of the God-of-Cosmic-Power invested in them. But we are not just physics. We exhibit the miracle of double comprehensibility, in that we can (in principle) be explained and understood personalistically, as well as physically. Personalistic explanation (1) cannot be reduced, even in principle, to physical explanation and (2) is intellectually fundamental. It is these two factors which give substance to the claim that *we* can be in control – it not being the case that ultimately it is just *physics* which decides. We exist. We possess the power to act, to create, to "disturb the universe", even though experiential physicalism is true. I have shown, at least, that this is *possible*.

Free Will and Wisdom

Instead of tackling the free will/physicalism problem, perhaps we ought to be tackling the wisdom/physicalism problem.[38] For wisdom presupposes free will, but free will does not presuppose wisdom. Hence the wisdom/physicalism problem is more general, more fundamental, than the free will/physicalism problem. The solution to the former automatically solves the latter, but not *vice versa*.

And there is another reason why we should, here, give priority to the wisdom/physicalism problem. Our fundamental problem in life, I have emphasized, is to help the God-of-Cosmic-Value to flourish within the God-of-Cosmic-Power – to help life of value to flourish in the real world. In so far as *we* have a role in this, what we require is *wisdom*, in the sense indicated in note 38. It better be possible, then, for us to have wisdom in the physicalistic universe, or our fundamental life problem becomes insoluble. How wisdom is possible granted physicalism is merely a preliminary to the problem that really matters: How can we *improve* what wisdom we already have? This preliminary problem is nevertheless more important and fundamental than the orthodox free will/physicalism problem.

But is the above right? Does wisdom presuppose free will? Someone might be wise in that they have the capacity and active desire to realize what is of value in life, and yet they might not have free will in that they pursue this aim in an obsessive,

---

[38] Remember, I am taking wisdom to be the capacity and active desire to realize what is of value in life for oneself and others: see note 1 of chapter 2.

compulsive way, helplessly in the grip of the pursuit of value, unable to do otherwise, imprisoned within the pursuit of this goal. A scientist, artist or politician might, for example, relentlessly and compulsively pursue a goal unquestionably of great value to humanity, achieve great success, and yet be lacking in free will because they cannot, in any circumstances, do other than pursue this narrow (but immensely valuable) goal.

But then we can also argue that this does not constitute authentic wisdom – or constitutes only imperfect wisdom. The authentic article would be such that one can adapt what one pursues in a wide range of circumstances, so much so that, even if what is of value is transformed by circumstances, one can adapt to these new conditions. Wisdom, in this sense, it may be argued, is a genuinely stronger notion than free will, in that anyone fully wise has free will, but plenty of people who have free will are not wise.

To spell this out in a little more detail, wisdom is a multi-dimensional notion. Our wisdom may be inadequate in many different ways. Our capacity to realize what is of value may be fragile or robust, restricted to one special kind of value, or broad in scope. It may be more, or less, selfish. It may be more, or less, skilful, knowledgeable, intelligent. It may have more, or less, desirable human qualities, such as humanity, friendliness, integrity, courage, sense of humour, empathy for others, creativity, originality. If a person who is actively realizing what is of value in life exhibits inflexibility, compulsiveness, lack of self-control, and thus lack of free will then, so the argument goes, this just reveals that the person is only inadequately wise. A person who can realize what is genuinely of value in a wide range of circumstances – to himself and others – cannot, to that extent, be lacking in free will. Any imaginary circumstance invented to reveal this person's lack of free will would, if successful, also reveal that, in these circumstances, the person cannot realize what is of value, and would thus demonstrate that person's lack of wisdom.[39]

---

[39] One could argue, of course, that free will is a value in its own right, and hence a person who does not have it cannot be wise. This is one way of arguing that one is only wise if one has free will. The argument in the text is rather different. It is that lack of free will implies inadequacy in the capacity to realize what is of value in life, and hence inadequate

Perhaps free will needs to be viewed in a new light, as a minimal requirement for, or specification of, wisdom. On the other hand, one may be inclined to dismiss this idea out of hand. Consider two brutal criminals, one who acts with full possession of free will, and the other who acts with much diminished free will. We would surely not want to say the former is *wiser* than the latter; if anything, we might be inclined to put the matter the other way round. But if free will is a step towards wisdom, we are, it seems, obliged to say the criminal who acts freely is wiser than the criminal who does not.

Perhaps the multi-dimensional character of wisdom can be invoked here. One way in which wisdom can be degraded is for "what is deemed to be of value" to be degraded. Another way is for one's self control to become degraded. The first criminal has become unwise in the first way, the second in the second way.

Be that as it may, it is worth noting that the notion of free will is linked to notions of value. First, free will is, surely, of value in itself, and not merely of value as a means to other things of value. Secondly, a person who pursues goals in life that have no conceivable meaning or value whatsoever would hardly be regarded as exhibiting free will. They would rather be regarded as mad. Thirdly, whether one holds a person to be pursuing a goal freely or obsessively (i.e. with diminished free will) may well depend on the value that is assigned to the goal. A person who devotes his entire life to surfing ceases to be obsessive if surfing is indeed taken to be the be and end of all life, and everything else is deemed pointless or irrelevant (unless subservient to surfing). Free will, like wisdom, in other words, is impregnated with value. Furthermore, in order for a person to have free will it is necessary, we may hold, that he can pursue goals of some minimal value to him to some minimal degree of success in a minimal range of circumstances. Put like that, free will begins to look like a minimal version of wisdom.

Reformulating the free will/physicalism problem as the wisdom/physicalism problem could be construed as favouring compatibilism. Free will puts the emphasis on responsibility,

wisdom.

choice, being able to do otherwise. The compatibilist account of these things seems to many to be seriously inadequate. By contrast, wisdom (as understood here) puts the emphasis on the pursuit of a goal – that which is of value. Goal pursuing in general – and this goal pursuing in particular – are much easier to understand as something that can go on in a physicalistic universe, as I have indicated above. What matters, granted that goal-pursuing is the issue – in particular pursuing the goal of what is of value in life – is responsibility, the power to choose, to do otherwise, in precisely the compatibilist's senses, since it is this we need in order successfully to realize our goal in a wide range of possible circumstances. It is not at all obvious that responsibility, choice, the power to do otherwise, in incompatibilist senses, are required for, or help with, successfully realizing a goal.

Consider moral responsibility – sometimes invoked in philosophical discussions of free will. A man dives into a dangerously fast-flowing river and saves a child from drowning. Was he acting in a morally responsible way? On this occasion TV cameras happened to be on the river bank recording the rescue. Did the man act out of vanity, in order to become a national hero, or did he act morally? In order to know the answer we need to know how we would have acted if there had been no TV cameras, and no possibility of becoming a hero. The crucial question, in other words, is how we would have acted in relevant, different circumstances. But this invokes the compatibilist notion of responsibility, of "could have done otherwise" (or, in this case, "could have done the same"), not the incompatibilist notion, which is hardly the issue. Compatibilism does better justice to moral responsibility than does incompatibilism.

Will Power

Free will is sometimes thought of in terms of *will power*. If our will is *powerful*, we have a lot of free will; if it is weak, we don't. I have myself flirted with this notion when I have declared that we have all the power of the God-of-Cosmic-Power with us when we decide upon and initiate our actions. But this traditional way of thinking of free will has perhaps become nowadays (among the non-religious) somewhat discredited, because of its past associations with Christianity.

Granted Christianity, the big problem in life is to do the will of God and resist temptations of the devil. This can degenerate into the problem of pursuing long-term interests (admittance to Heaven) and resisting short-term pleasures of the flesh – food, drink, and above all, of course, sex. Free will thus becomes the power to resist short-term desires in the interest of pursuing and realizing long-term aims. The very idea of the *power* of the will may have arisen in this way. We have a powerful will if we can resist short-term temptations and pursue successfully long-term aims, and we have a weak will if we pathetically give in to short-term temptations.

This idea of will power is, however, by no means restricted to Christianity. It is meaningful in many other contexts. For many other doctrines and philosophies of life hold that goals of value are long-term ones, a fundamental problem of life being not to give in to short-term temptations, thus allowing long-term goals to be compromised or lost. This is true of other religions, such as Judaism and Islam. It is true of the capitalist idea that what is of supreme value in life is wealth, virtue residing in the life-long pursuit of wealth, short-term pleasures and distractions being ignored. It is true of the idea that what matters in life is long-term reputation, whether moral, scientific, literary, artistic, or political. Some may even hold that wisdom itself constitutes such a goal, wisdom being acquired gradually over a life-time, but only if short-term distractions are resisted. All such doctrines hold that a fundamental problem is to hold onto long-term goals of value, and to resist short-term temptations: all these doctrines find a role for will power, even if somewhat differently interpreted in detail.

We may, however, hold that all such philosophies of life, which insist that short-term temptations must be resisted for the sake of long-term goals of value, have got it wrong. Much of the value of life resides in what is immediate and ephemeral – in just those short-term pleasures and joys that Christians and other Puritans condemn. We may even hold that those who successfully exercise "will power", in the interests of achieving long-term goals, are actually very seriously lacking in free will. Such people are obsessed, imprisoned in a false philosophy of life that condemns them to an impoverished way of life and abjure much that is of value in life. They are in thrall to the "power" of the will, imprisoned in it. From this perspective, will power seems very different from free will.

We can, however, I think, reinterpret the notion of will power so that it does not have these unfortunate connotations. At the same time we can, I think, learn something important from what has emerged from the discussion above, namely that there may be a link between how we conceive of free will, and what we take to be the fundamental problems of life.

It is clearly wrong to think that will power invariably involves overcoming short-term temptations in the interests of long-term goals. The exercise of free will, and therefore will power, may well involve, on occasions, allowing short-term temptations to supplant, or defer, long-term goals. For free will in the best sense, what we need to be able to do is resolve *conflicts* between clashing desires, aims, interests, so that what is genuinely of value, genuinely in our best interests, is achieved. Will power needs to be reinterpreted to mean: the power to resolve conflicts so that what is genuinely of value may be achieved. This fits in well, of course, with the idea that free will should be interpreted as the capacity to realize what is of value in life – i.e. that it should be interpreted to mean "wisdom".

The traditional notion of will power, associated with Christianity, is thus correct to interpret will power to be the power to resolve life's fundamental conflicts, but wrong in its Christian view as to what these conflicts are, and wrong in its view as to what constitutes a good resolution of these conflicts. Granted that the Christian view on these matters needs to be rejected, the question arises: what do we put in its place? What are life's fundamental problems and conflicts, and what constitutes a good resolution of them? What we take will power to be will depend, to some extent, on how we answer this question.

At the most fundamental level of all, the answer of this book is quite clear. Our fundamental problem in life is to realize what is of value, potentially and actually, in the circumstances of our life. It

259

is to realize what is of value, enmeshed as we are in the physical universe.

The next chapter on evolution will suggest some rather more specific conflicts we all inherit because of our evolutionary past.

First, as I have indicated above, animal life is riven with conflict, often involving life and death decisions. The mouse must leave the security of its nest to look for something to eat, but in doing so risks itself becoming a meal for an owl. Survival and reproductive success depends crucially on having the power to resolve such conflicts successfully. That mouse survives and reproduces which can successfully resolve the conflict between the need for safety, and the need for food. Granted we interpret will power to be the power to resolve conflicts in life, it is clear that Darwinian theory can account for the evolution of will power.

Second, what Darwinian theory accounts for is the power to resolve conflicts and problems of living *so that survival and reproduction may be achieved.* This is what we inherit, as it were, from our evolutionary past. But what we require for our human will power is the power to resolve conflicts and problems of living so as to realize what is of value – which may include, but is not the same as, survival and reproduction. A fundamental life problem, then, is to convert the Darwinian aims (as we may call them) of survival and reproduction into aims of realizing what is of value. I will have more to say about this in the next chapter.

Third, a basic conflict we inherit from our evolutionary past can be conceived in the following terms. We attempt consciously to plan our way of life, but animals do not. Animals may be said to plan consciously what they do from moment to moment, but a master control system of hormones determines the way of life by prompting the animal to hunt, sleep, mate, fight, care for young, etc., at appropriate times and in appropriate contexts. This "master control system" exists in us too but, because our conscious awareness has enormously increased over that possessed by animals (as a result, I shall argue, of the development of language and the ability to imagine), we try to plan consciously our way of life. This puts our conscious planning in direct conflict with our evolutionary "master control system". Here is the source of a conflict likely to be fundamental for all human beings – a distorted echo of which is to be found in the Christian view as to what our basic life conflict is. I shall say more about this conflict in the next chapter.

Human life provides far more conflicts than does animal life (even if they do not usually involve the life and death decisions of animal life). This is because of the far greater range of choices of value that arise in the human context. A more detailed discussion of will power might explore some of the myriad conflicts and problems of living we encounter in life, and how they are to be best resolved.

In conclusion, then, we can say this. We need to rescue the notion of will power from its past Christian or Protestant associations. If we do so, and interpret it as the power to resolve conflicts, it becomes an entirely legitimate notion. Will power in this sense can readily be seen to be something that will evolve as a result of the twin Darwinian mechanisms of inheritable variations and natural selection. In order to clarify and give substance to what we mean by will power, we need to specify what our fundamental conflicts and problems of living are, and how they are best to be resolved.

The Miracle of Free Will

I have argued that free will is *possible* even though physicalism is true. But one has to add that this argument renders free will little short of a miracle. For what is required is that the God-of-Cosmic-power – the minute bit of It that is us – must be so extraordinarily and intricately organized that successive physical states, unfolding in time in accordance with iron physical law, just happen to be us consciously acting in the world, deciding what we are going to do and then doing it, actually producing and guiding our actions. The God-of-Cosmic-Power becomes our servant, doing our bidding, evolving in time so as to *be* us, we coherently in charge with the flow of our thinking and acting. We are doubly comprehensible – simultaneously comprehensible physically, and personalistically. It is this double comprehensibility that is the miracle. How can it have come into existence? How can we explain and understand its existence? It is this crucial question that I tackle in the next chapter.

261

Do We Have Free Will?

All this intricate philosophising is all very well – the cry may go up – but the all important question is: Do we have free will? A famous experiment by B. Libet (1985) has been taken by many to establish that we do not. Subjects of the experiment were asked to move their hand at an arbitrary moment decided by them, and to say when they made the decision, determined by noticing the position of a dot going round a clock face. Monitoring of the electrical activity of the subjects' brains established that the Readiness Potential – a pattern of electrical activity known to precede voluntary action – occurred some tenths of a second *before* the conscious decision to move the hand. This has been taken to demonstrate the illusory nature of free will, since it is the brain that causes both the hand to move and the subsequent conscious "decision". For free will, we would require that the conscious decision causes the Readiness Potential to occur subsequently in the brain which in turn causes the hand to move.

But the conclusion does not follow. The free action here, we may argue, is the decision to move the hand at some future arbitrary moment, as required by the experiment, which is, in effect, the decision to put the brain into a special state such that fluctuations in brain activity of a sufficient size lead to the hand moving. That the subsequent decision to move the hand now, at this arbitrarily chosen moment, is not free does not undermine the reality of free will in general, nor the freedom of the decision to take part in the experiment, and put one's brain into the special state required. Most real-life decision-making takes time, and rarely involves picking some moment instantaneously and arbitrarily.

Elsewhere,[40] I have put forward the following argument in support of the reality free will. Accept that, in order to achieve what is genuinely of value, we must have free will. If physics has made real progress in revealing to us the nature of the physical universe, then we have achieved something genuinely of value, and hence, by our assumption, we have free will. If, on the other hand, physics has not made real progress, then the only reason for

---

[40] See my (1984), p. 274, or (2007a), p. 295.

doubting the reality of free will collapses – for that stems from the conception of the universe that has emerged from physics which, in this case, is false. Either way, we should conclude free will is real (or at least that there are no good reasons to doubt its existence).

When it comes down to it, perhaps we should invoke a revised version of Descartes' "I think, therefore I am" to establish the reality of free will – or at least that we are entitled to accept that free will is a reality. What is at issue, as I said at the beginning of this chapter, is our existence. If we do not have free will, we do not exist. We are rationally entitled to hold that we do have some free will because in making this assumption we have nothing to lose and everything to gain. We should not allow ourselves to be bamboozled into thinking we do not exist. I exist, therefore I have free will.

There is an analogy, here, perhaps, with the problem of knowledge. Traditionally, this has been taken to be the problem of how we can acquire any knowledge at all. Popper, in my view correctly, transformed this fundamental problem of epistemology so that it becomes "How can we acquire *more* knowledge?" or "How can we *improve* our knowledge?". It may seem illegitimate just to presuppose at the outset that we do already have some knowledge, but without that presupposition we are lost. We are rationally entitled to assume that we do have some knowledge, and the fundamental task is to improve what we have, because if we have no knowledge we have no basis for acquiring it. I suggest something similar holds for the problem of free will. The fundamental problem is not "Do we have free will?" or "Can we have free will in a physicalistic universe?". Rather it ought to be "How can we best go about *increasing* or *improving* what free will we do already possess?". An important part of the answer is to put aim-oriented rationality and wisdom-inquiry into practice in our lives.

CHAPTER EIGHT

EVOLUTION OF LIFE OF VALUE

We saw in the last chapter that free will is possible but wildly implausible granted physicalism. Whatever else we may be, we are at least a bit of the physical universe – a fragment of the God-of-Cosmic-Power. It is just about possible that this bit of the physical universe in which we have our being – our brains, bodies and environment – is so beautifully and intricately convoluted, structured, organized and designed that physical law, unfolding in its remorseless, unthinking way, just happens to be also us, freely deciding what to do, and then making what we have decided happen. It is just conceivable that we, and the God-of-Cosmic-Power, have dual control – what we are being doubly comprehensible. We just about could have all the might of the God-of-Cosmic-Power within us, so utterly devoted to our interests as to empower us to initiate and guide our actions.

But if so, this state of affairs really is wildly, incredibly implausible, little short of the utterly miraculous. Why should the fragment of the utterly impersonal physical universe we inhabit be so intricately and conveniently designed and organized so as to facilitate us being in charge of our thoughts, decisions, and actions (at least some of the time, to some extent)? This seems utterly inexplicable. It cries out for explanation.[1]

This profound problem of explanation and understanding has been solved, in outline at least, by Charles Darwin. The solution is his theory of evolution. The blind, purposeless mechanisms of random inherited variations and natural selection, operating initially on some elementary, initial life form have, during the course of some three and a half billion years, produced the millions of diverse living things that inhabit the earth today, including ourselves, all incredibly well-adapted to their conditions of life,

---

[1] Incompatibilists may hold their view in part because they see compatibilism as being untenable because, at best, it has this apparently inexplicable, absurd consequence.

and so able to survive and reproduce. The blind mechanisms of evolution design both bodies and brains. As a result of designing brains, these mechanisms of evolution build into brains the capacity successfully to pursue those goals in the given environment that are conducive to survival and reproduction, plus the capacity to learn. The eventual outcome has been us human beings, imbued with the capacity to decide for ourselves (some of the time, to some extent) what we want, what we will do, plus the capacity to do it. The miracle of free will is, in other words, the outcome of Darwinian evolution.

However, if Darwinian evolution is to explain the miracle of the existence of free will in this physicalistic universe, it is essential that we adopt that version of Darwinian theory able to perform this task. In what follows, I shall distinguish eight versions of Darwinian theory. Only the final, eighth version is able to explain free will, as we shall see.

Actually, the task before us is broader than to account for free will in the universe. Our fundamental problem is to understand how the God-of-Cosmic-Value has evolved within the God-of-Cosmic-Power. How has all that is of value emerged within the physical universe? I concentrate on a key component of this problem, namely the evolution of the *capacity* to realize what is of value. This capacity may be called *wisdom*, and wisdom, as we saw in the last chapter includes, but goes beyond, free will. I set out to answer two key questions: (1) What version of Darwinian theory is able to explain the evolution of wisdom? (2) How good, how adequate, is this explanation? What are its limits, its inadequacies? Understanding how wisdom (in the sense indicated) has evolved is crucial to understanding how the God-of-Value has evolved within the God-of-Cosmic-Power.

Nine Versions of Darwinian Theory
The task before us is to specify a version of Darwinian theory which provides the best available explanation for the existence of human beings who are *doubly comprehensible* – comprehensible *physically*, and comprehensible *personalistically*. We want to understand how beings have come into existence in the physical

universe who are *able freely to realize what is of value in life* (at least some of the time, to some extent).

Formulating the problem in this way, as understanding how beings that are *double comprehensible* have come into existence, makes it clear that the sought for explanation must itself take account of *both* kinds of explanation – physical and personalistic.

Darwinian theory is a very special kind of *historical* theory. All historical explanations – including Darwinian ones – make use of other modes of explanation, such as the three discussed in the previous chapter: physical, purposive, and personalistic. But in the case of Darwinian theory, the appeal to these other modes of explanation arises for a much more basic reason. The theory seeks to understand how and why things exist – living beings – that are amenable to being explained and understood simultaneously in two (or even three) different ways: physically, purposively and, in some cases, personalistically. This can hardly be achieved if these modes of explanation are ignored.

Darwinian theory is thus, on this view, quite different from Newtonian theory say, or most other scientific theories, which seek to predict and explain a range of phenomena, but which do not seek to explain why some things are doubly (or in some cases trebly) comprehensible. Unlike other scientific theories, the problem for Darwinian theory is not the incomprehensibility of a range of phenomena, but rather that some phenomena – having to do with life – are, as it were, *much too comprehensible*, in being *doubly or even trebly comprehensible*. It is the excessive comprehensibility of life that is the problem.

Darwinian theory solves this problem historically, by explaining how and why increasingly diverse and rich double (and eventually treble) comprehensibility has come gradually into existence over billions of years in an initially purposeless, singly comprehensible universe. This problem can only be solved in this way, however, if Darwinian theory observes the following principle:

Principle of Non-Circularity: The theory must not presuppose what it seeks to explain. If, at some stage in evolution, Darwinian theory itself employs purposive explanations, the theory must explain how purposiveness of this type has come into existence at this stage of evolution *without using the very notion of purposiveness that is being explained.* And just the same applies to the personalistic.

This Principle must be observed if Darwinian explanations are to avoid becoming trivially circular – presupposing the very thing to be explained. Darwinian accounts of evolution may employ purposiveness and personalistic explanations, at certain stages of evolution, but if so, Darwinian theory must explain how things that exemplify these notions of the purposive or personalistic have come into existence *in a way which makes no appeal to these explanatory notions whatsoever.* Thus, if an appeal is made to empathy in order to explain some evolutionary development, an explanation for the prior evolution of empathy must be given *which does not itself employ empathy as an explanatory notion.* Or, if parental care is employed to explain some evolutionary development, the existence of parental care must itself be explained without this explanation itself invoking parental care. And likewise for purpose, sentience, consciousness, free will, cooperativeness, and so on.

If this Principle is observed, we have a theory which may be able to explain the emergence of the purposive and personalistic in a purposeless universe; if it is violated, the whole programme collapses. Darwinian theory merely presupposes what it sets out to explain.

We shall see that Darwinian theory is at present only partially successful in conforming to this Principle of Non-Circularity. One difficulty arises in connection with the unsolved problem of the origin of life.

I now consider eight versions of Darwinian theory which, progressively, give increasingly important roles to purposive and personalistic modes of explanation.[2] I begin with an extreme

---

[2] The account of Darwinian theory developed here, stressing that the theory needs to be interpreted as explaining double (or treble) comprehensibility, this requiring that the theory itself appeals to purposive and personalistic modes of explanation, the mechanisms of evolution themselves evolving, is based on a much more detailed exposition of all this in my (2001), ch. 7.

version of the theory that banishes "purpose" from the theory (and from life) altogether. I do this so that we may have before us the full range options, from the extreme mechanistic and purposeless on the one hand, to the fully personalistic on the other. The first, purposeless version might be attributed to Jacques Monot and Richard Dawkins. Let us call it:

Darwin(1) The theory is about the evolution, not primarily of living things, but rather of entities that may be called *replicators*. These are genes, encoded in DNA molecules. Replicators replicate themselves by manipulating the "survival machines", or bodies, they inhabit. Evolution of replicators occurs as a result, in essence, of (1) random inherited variation (mistakes in the process of replication), and (2) the natural selection of those replicators best able to survive and replicate.

Comments. This seems to invoke purpose, in that replicators are described as performing such purposive actions as *replicating* themselves by *manipulating* their survival machines. Upholders of this view would insist, however, that this is just convenient metaphor. All reference to purposive action can be eliminated from the theory.

Does anyone defend Darwin(1)? Dawkins certainly seems to, in his *The Selfish Gene*.[3] At one point he says "[The genes] are the replicators and we are their survival machines" (p. 37), and this theme is spelled out at some length in the book. "... the fundamental unit of selection, and therefore of self-interest, is not the species, nor the group, nor even, strictly, the individual. It is the gene, the unit of heredity" (p. 12). He even says at one point that these replicators "are in you and me; they created us, body and mind; and their preservation is the ultimate rationale for our existence" (p. 21). And Dawkins states explicitly that it is quite wrong to invoke *purpose*. He says "natural selection favours replicators which are good at building survival machines, genes which are skilled in the art of controlling embryonic development. In this, the replicators are no more conscious or purposeful than they ever were. The same old processes of automatic selection

---

[3] Dawkins (1978).

268

between rival molecules by reason of their longevity, fecundity, and copying-fidelity, still goes on as blindly and inevitably as they did in the far-off days. Genes have no foresight. They do not plan ahead. Genes just *are*, some genes more or so than others, and that is all there is to it" (p. 25).[4]

It is possible to interpret Darwinian evolution in this way, but it seems bizarre and perverse to do so in the extreme. It is as if what one finds utterly amazing and in need of explanation is not life on earth – plants, fish, birds, mammals, human beings, in all their extraordinary diversity, living their extraordinarily diverse ways of life – but DNA molecules. (I once heard Richard Dawkins begin a

---

[4] The gene-centred view is very clearly expressed and defended by Helena Cronin (1991). She writes "Modern Darwinian theory is about genes and their phenotypic effects. Genes do not present themselves naked to the scrutiny of natural selection. They present tails, fur, muscles, shells; they present the ability to run fast, to be well camouflaged, to attract a mate, to build a good nest. Differences in genes give rise to differences in these phenotypic effects. Natural selection acts on the phenotypic differences and thereby on genes. Thus genes come to be represented in successive generations in proportion to the selective value of their phenotypic effects" (p. 60). And she adds "We have travelled far from the organism-centred view of classical Darwinism – from a Darwinism that is about the survival and reproduction of individuals" (p. 64). She goes on to stress the importance of strategic thinking in modern Darwinism, and adds "The development of strategic thinking has involved two major shifts from classical Darwinism: first, a view of adaptations that is more conscious of their costs and less sanguine about their benefits, and, second, a greater emphasis on behaviour, particularly social behaviour. The strategists, of course, are not runners – not even robins or rats: they are genes" (p. 66). At first it almost sounds as if modern Darwinism takes purposiveness very seriously indeed, in emphasizing strategy and behaviour, especially social behaviour. But then all this is removed with the declaration that the strategists are genes! For of course genes can only be said to be selfish, strategists or, more generally, purposive, in a metaphorical, not in a literal sense. Only living things are purposive. In re-interpreting the theory of evolution to be about genes rather than living things, modern Darwinism, almost incidentally, perhaps entirely unintentionally, removes purposiveness from the theory altogether – or, at the very least, downplays its role in evolution.

lecture at University College London with the words "My vision is a world full of replicators"!)

Why does Dawkins take the unit of selection to be the gene, the replicator, and not the individual living thing – the "survival machine" to use his term? Because genes endure thousands, even millions of years, individual exemplifications of a given gene are precisely replicated, and the gene is invariably selfish. Individuals, by contrast, have a short life; they are all different, do not reproduce precisely, and are not invariably selfish (in that they are sometimes altruistic, as when bees sting animals after honey, and so die to save the hive). But these differences do not seem to me to constitute any argument at all against:

Darwin(2) The theory is about the evolution of individual living things – bacteria, viruses, fish, insects, birds, reptiles, mammals, plants, fungi and the rest. These have evolved, and continue to evolve, as a result, in essence, of the twin mechanisms of random inherited variations and natural selection. Living things appear to pursue goals, but they don't really. What the theory does is to explain away the illusion of purposiveness in nature. Life is just a combination of chance and necessity, devoid of purpose.

Comments. Many biologists do, or have, accepted Darwin(2), although many others reject it. Dawkins' reasons for preferring Darwin(1) to Darwin(2) do not seem to amount to very much. Why should the unit of selection persist for thousands of years? Why should reproduction precisely replicate what is reproduced? Darwinian theory is about reproduction *with variation*. Even the argument that only genes are always selfish does not seem to amount to much. Altruistic action undertaken to save close kin may be thought of as engaging in a kind of reproduction. One reproduces, not by having offspring oneself, but by protecting the lives of close relatives likely to have offspring of their own. Thus, all that needs to be done in order to make such altruistic action in no way exceptional, but a standard part of action designed to promote survival and reproductive success, is to broaden the meaning of "reproduction" a bit. This is something one needs to do for other reasons in any case, as we shall see below.[5]

The substantial reasons for preferring Darwin(2) to Darwin(1) only really arise, however, when we come to consider versions of Darwinian theory that attribute genuine *purposes* to living things. Whereas it makes sense to hold that living things pursue goals, it does not make quite so much sense to think of genes, stretches of DNA molecules, as genuinely engaged in purposive activity. It may well be that a part of Dawkins' reason for preferring Darwin(1) to Darwin(2) lies in just this feature of the former view – its clear mechanistic, purposeless character, as his talk of *replicators* and *survival machines* suggests.

But does Dawkins' really deny that purpose has anything to do with evolution? It is an awkward denial, for two reasons. First, it creates a wholly artificial division between humanity, authentically purposive in character, and the rest of the living world, devoid of purposiveness according to Darwin(1) and Darwin(2). This problem – this gulf between humanity and the rest of the living world – so much against the whole spirit of Darwinianism, which is all about gradual evolution – is merely an artefact of the above

---

[5] The worker bee or ant, sacrificing itself in order to save the hive or its close kin, is an extreme case of something less extreme and much more widespread, namely parental care. This involves some self-sacrifice in order to promote the survival of one's offspring, even if not the supreme self-sacrifice of one's own death – although parental care may go to that extreme, when predators are fought or distracted to preserve the lives of offspring, for example. Parental care involves acting so as to promote the survival of one's own offspring, whereas the sacrifice of worker bees or ants promotes the survival of offspring of near relatives: somewhat different, but not fundamentally different. In this context, one should perhaps remember that there are many others cases of living things participating in reproduction even though their own genes are not reproduced. One might think, for example, of bees fertilizing flowers and blossom by distributing pollen, or birds and mammals eating fruit and thus distributing seeds. In these cases, of course, the bees, birds and mammals are after food; they are induced by the reward to serve, unknowingly, the reproductive needs of another species. Nevertheless, it is worth keeping in mind the wonderful variety of activities associated with reproduction.

two versions of Darwinian theory, perversely denying purposiveness to non-human living things. Second, Dawkins, like other biologists, is prepared to talk of *design*. But the notion of design presupposes the notion of purpose. Whether something is well or ill designed may well depend crucially on what purpose it is being considered for. A chair that is well-designed as an object to be sat in is appallingly designed if considered to have the purpose of a teaspoon – to scoop up jam or stir sugar into one's tea.

But Dawkins' denial of purpose is perhaps a somewhat trivial semantic matter, rather than a matter of substance. In *The Selfish Gene*, Dawkins makes clear that he takes purposiveness to mean "conscious purposiveness", and he goes on to say that machines, such as guided missiles and computers playing chess, can be made to act *as if* pursuing goals by means of feedback mechanisms and computer programmes (p. 53-6). If one restricts oneself to a narrow notion of purposiveness – one that requires all purposes to be conscious, or one that insists the actions of the thing in question cannot even in principle be explained physically – then one will be forced to deny purpose (in these narrow senses) to living things. Broaden the meaning of "purpose" so that it becomes free of these restrictions, and becomes such that it includes the compatibilist notion explicated in the last chapter, and it becomes utterly absurd to deny purposiveness to living things. Dawkins himself, indeed, would agree (although, perversely, purposes are attributed by him in the first instance to genes, to sections of DNA molecules, rather than to living things themselves). This brings us to:

Darwin(3). Living things are inherently purposive beings. Their fundamental goal in life is survival and reproductive success, and all their other goals contribute, in one way or another, more or less successfully, to this fixed, fundamental goal. The *mechanisms* of evolution are, however, blind and purposeless.
Comments. On this view, Darwinian theory does not explain purpose away. On the contrary, it explains how purposiveness has gradually crept into Nature.[6]

---

[6] See Maxwell (1984), pp. 174-181 and 269-275; or Maxwell (2007a),

It is probable that most biologists uphold Darwin(3). Those who reject the idea that living things are purposive probably do so for reasons similar to Dawkins'; they interpret "purpose" much too narrowly, to mean either "conscious purpose", or "purposiveness that is incompatible with physics".

Darwin(3) is however untenable because, once it is admitted that animals pursue goals, it becomes impossible to keep the mechanisms of evolution free of all elements of purposiveness, as we shall now see.

Darwin(4). Not only are living things purposive. The mechanisms of evolution, inherited variation and natural selection, themselves evolve, incorporating, as they do so, elements of purposiveness – so that these mechanisms can no longer be described as purpose-free. Animals in effect *breed* other species, or even their own species, by their purposive actions, even though they are not aware, of course, of what they are doing. To say this does not mean, however, that evolution itself has a purpose.

Comments. There are at least four ways in which purposiveness insinuates itself into the mechanisms of evolution.

(a) What has survival value may depend on how the animal is living. A change in the way of life may be due to a change of habitat, or a change in the climate. Animals may even create their habitat, as beavers do when they build dams and create lakes. In order to explain subsequent evolutionary developments it will be necessary to refer to prior purposive action, and prior changes in purposive action, among other factors. Whether a mutation has survival value or not will depend on how the animal is living. For a dog-like creature running about on land, a mutation which turns legs into flippers would be a disaster. But if the creature is already swimming in rivers, catching fish, this mutation would have immense survival value.

(b) Sexual selection. One sex – typically females – may prefer to mate with those who possess certain characteristic features, and as a result, those features will tend to become more prevalent and

---

pp. 197-205 and 290-296. This "purposive" version of Darwinian theory is further elaborated in Maxwell (2001), ch. 7. See also Maxwell (1985).

exaggerated in the population. Thus, peahens, choosing to mate with those peacocks with the most splendid tails, inadvertently partly cause peacocks to have absurdly splendid tails. In order to explain the peacock tails, it is necessary to refer to the purposive actions of peahens, amongst other factors.

(c) Offspring selection. Parents, in choosing preferentially to feed offspring with certain characteristics – allowing offspring who do not have these characteristics to die – may thereby be a part of the cause of these characteristics to become more prevalent in the population. Likewise, some offspring may be better at manipulating parents to feed them than others, thus increasing the likelihood of their survival, and the spread, through the population, of these genetically determined manipulative techniques.

(d) Predator-prey selection. The fox, in hunting rabbits, kills those rabbits not so good at evading capture and death. In this way the fox helps breed rabbits better and better at evading foxes. And likewise, rabbits, in escaping from those foxes not so good at hunting, help breed foxes better and better at hunting (since foxes not so good at hunting tend not to survive and reproduce). Similarly, birds breed caterpillars and butterflies good at camouflage, and the latter, in getting better and better at camouflage, may help breed more perceptive birds able to see through it. Yet again, plant eating animals may help breed plants better able to resist the destructive attention of the animal in question. And the plant may help breed animals better able to eat the plant.

If we grant that plants engage in purposive action in the main by means of *growth*, then we may extend the idea that purposive action influences evolution from animals to plants. Plant growth creates soil, and creates shade, both of which have had consequences for subsequent evolution. Shade in tropical rain forests creates selective pressures for young plants, in a clearing where there is sunlight, to grow quickly so as not to fall into the shade of more quickly growing plants. It also creates selective pressures for plants that can do photosynthesis in shady conditions. Genes that generate these traits will be selected for. And of course those cells, early on in evolution which, as a result of growth and

274

photosynthesis, generated oxygen in the atmosphere made possible all of animal life.

In the above four kinds of cases, actions of animals (or growth of plants, in itself a kind of purposive action), has an impact, along with other factors such as mutations, on subsequent evolution. In all four cases, what is involved may be called *breeding*, analogous to what pigeon fanciers and dog owners do when they breed new varieties of pigeon or dog, with the crucial difference that the animals, or plants, have no idea whatsoever of what they are doing. Their purposive actions have the effects of breeding without there being anything like the conscious purpose of breeding. But then human beings may engage in breeding without being aware of what they are doing. In the case of (a) and (c), the animal unconsciously self-breeds; that is, it breeds its own offspring, its own species.

In *The Human World in the Physical Universe* I suggested another slogan for evolution, to take into account these phenomena: *life breeds itself into existence!*[7]

All this, it should be noted, is wholly orthodox Darwinian theory. Selection still acts on individuals. It is just that how it acts depends to some extent on how the animal acts, what kind of goals it pursues; and the environment in which an animal lives consists in part of other living things whose actions impact on the given animal. Selection is not entirely blind; there is a purposive element, even if no foreseeing of what the outcome will be.

Nevertheless, the transition from Darwin(3) to Darwin(4) makes a profound difference, in my opinion, to the way one should view evolution. For the latter version of the theory, unlike the former, makes the actions of animals, our ancestors, a vital part of the explanation of our existence. Evolution is not just blind chance and necessity. Our animal ancestors, striving to live, to eat, to avoid being eaten, to mate, to rear young, are a vital part of the reason for our existence. They did not, of course, intend us to exist. Nevertheless, without their striving, we would not be here. We owe them a debt of gratitude.

---

[7] See Maxwell (2001), p. 174.

Darwin(5). Once individual learning, and the capacity to imitate, have come into existence, evolution by cultural means[8] becomes possible – a kind of evolution that mimics Lamarckism, in that acquired characteristics are (culturally) inherited. An individual learns to do something new, others imitate the action, and it becomes a persistent activity of the group, even though no genetic changes have taken place. Purposiveness has become a part of the mechanisms of evolution in a much more radical way. These mechanisms have themselves evolved in a much more substantial fashion.

Comments. As an example of evolution by cultural means, one might take the very well-known case of chimpanzees eating termites. An individual chimpanzee discovered that, by pushing a stick into a termite nest, leaving it there for a bit and then withdrawing it, termites, clinging to the stick, can be eaten off it. (Chimpanzees may have started by sticking fingers into termite nests, and then learnt that sticks serve better.) Other chimpanzees, imitating what this one chimpanzee does, learn to do likewise. The trick is then passed on, via imitation, to offspring (and others). This is known to have emerged as a result of evolution by cultural means, and not as a result of some genetic change.

Evolution by cultural means is best understood as the development of a new method of reproduction. The characteristic way of life is reproduced, in part by the standard genetic means of sex, embryological development, birth and growth, but also, in part, by means of imitation. Reproduction by imitation makes possible quasi-Lamarckian evolution. An acquired characteristic – a new kind of action conducive to survival, learnt by an individual – can be passed on, by imitative reproduction, to offspring (and of course to others and their offspring).

---

[8] I employ the somewhat clumsy phrase "evolution by cultural means", and not "cultural evolution", because of the ambiguity of the latter. "Cultural evolution" might mean "evolution by cultural means", but in the relevant literature is generally taken to mean "the evolution of culture". Whereas "the evolution of culture" is about the evolution of a specific kind of thing – culture – "evolution by cultural means" refers to a specific manner in which evolution can proceed – by means of individual learning and imitation (or learning from others).

In order to construe evolution by cultural means as involving a new, or additional, method of reproduction, it is essential to interpret the theory of evolution as being about *life*, ways of living, and not as being just about bodies – let alone genes or DNA molecules. But this is, I maintain, the proper way to construe the theory in any case. Certainly if the concern is to understand how human life has come to exist, this is the proper way to interpret the theory. Once Darwinian theory is interpreted as being about evolving *characteristic ways of life* (including bodies and genes as an integral part of a way of life), it becomes inevitable that evolution by cultural means is to be construed as the development of an additional *method of reproduction* (superimposed upon genetic reproduction). For it is just that: a new way in which a bit of a characteristic way of life (eating termites off a stick) can be passed onto offspring and others.

Evolution by cultural means requires that individual learning, and the instinct to imitate, already exist. If the Principle of Non-Circularity is to be observed, an evolutionary account of the development of these capacities must be forthcoming which does not presuppose these capacities, let alone evolution by cultural means itself. Why, then, should the capacity to learn, and to imitate, have survival value (and therefore be selected when appropriate mutations arise)? The capacity to learn quite clearly has survival value. Even a primitive organism such as a sea anemone, with only a simple neuronal net for a brain, can learn. But what of the instinct to imitate? I suggest this has survival value, and is likely to have evolved, when there is parental care. Parents, just because they are parents, are likely to be good at survival and reproduction. Therefore, what they do is likely to have survival value. Hence, offspring imitating what they do is likely to have survival value. Thus, whenever there is parental care, and successful parents are around to be imitated, the instinct to imitate is likely to have survival value, as far as offspring are concerned.

Parental care is very ancient. Crocodiles, ancient beasts, engage in a form of parental care. It seems likely that dinosaurs did as well. So it may be that evolution by cultural means has its roots deep into our evolutionary past, well over 65 million years ago, and long before human beings existed.

Evolution by cultural means introduces an even more substantial element of purposiveness into the mechanisms of evolution. These mechanisms consist, in essence, as I have said, of two elements: (i) reproduction, with some inherited variation, and (ii) natural selection. The transition from Darwin(3) to Darwin(4) affects these mechanisms in affecting (ii), natural selection. This acquires some elements of purposiveness, as we have seen, even if it is not itself purposive, in that the outcome is not sought for, planned or intended. Evolution by cultural means introduces a more radical kind of purposiveness into the mechanisms of evolution by affecting (i) reproduction. The way of life is reproduced (with variation) in part by means of individual discovery and imitation. Both discovery, and imitation, are purposive (as understood in this context).

It may even be that the outcome is in part purposively intended. Cats and tigers teach offspring to hunt. Their actions may have the purpose of getting offspring to learn how to hunt skilfully and successfully.[9] Even in the pre-human, animal world, the outcome of elements of cultural reproduction may be purposively intended. If so, purposiveness here becomes an integral part of the mechanisms of evolution in a really substantial way.

Even though evolution by cultural means began long before human beings came into existence, it is above all with human beings that this form of evolution really comes into its own. As a species, we are very similar to others in all sorts of ways. We share 98.4% of our genes with chimpanzees. But in one dramatic way we are utterly unique. We are the product of evolution by cultural means to an extent far, far beyond anything found in any other species. It is this which accounts for the multitude of differences between us and all other species. Above all, of course, language is a product of evolution by cultural means. And language then makes endless other things possible, inaccessible to

---

[9] What is at issue, here, is not whether the mother cat consciously knows what she is doing, but whether her actions have as their goal (whether consciously or not) to teach the kittens to hunt well.

all other species. Art, science, democracy, justice, elaborate technology, planned social progress, even wisdom: these all become possible once there is language.

It is important to appreciate that evolution by cultural means, even though not itself involving genetic changes, may have an important impact on subsequent genetically determined changes. Consider again the dog-like creature running around, and hunting, on land. Suppose now that one individual, perhaps by accident, discovers that fish can be caught in a river, which are good to eat. Others learn by imitation. Many dogs spend time in the river hunting fish. Now a mutation appears making legs somewhat flipper-like. Given the new way of life, which has evolved culturally, flippers have great survival value, even though they would have been disastrous before the evolution by cultural means took place. The dog-like creature becomes a beaver-like creature, and evolution by cultural means led the way. It is a part of the reason for the evolution, from dog to beaver.

In reality, of course, such an evolutionary change would happen gradually, as a result of a combination of cultural and genetic changes, interacting with one another, over a long period of time. The really important point is that evolution by cultural means can have an impact on, can be an integral part of, genetically based evolution, the one intertwined with the other.

Almost certainly this took place in connection with the evolution of language. It seems reasonable to suppose that an elemental language came into existence first, perhaps by evolution by cultural means. Chimpanzees have three words in their vocabulary, one for snake, another for tiger. Once a primitive language exists, one can easily imagine that selective pressures exist for being good at speaking and understanding language. Perhaps this is required to mate, and have offspring. Perhaps men have more mates if they are good at speaking. Random genetic changes that produced brains, muscles and larynxes good at speaking would, in these circumstances, be selected for. Our human capacity for language would have evolved by means of an intricate interweaving of cultural and genetic developments, often called gene-culture *coevolution*.

279

A small but telling example of such coevolution in humans is cited Boyd and Richerson (2005), pp. 191-192. Most of the world's adults can't digest milk. Infants can but adults cannot. They lack an enzyme required to digest lactose, the sugar in milk. However, in those regions that have long had a history of keeping cows and dairying – Europe, parts of Africa and Asia – most adults can digest milk. The ability to drink milk is due to a single gene, widespread in those areas that have a history of dairying. As a result of learning to keep cows – an example of cultural evolution – the gene for digesting milk has survival value, and spreads in dairying populations, something it does not do in populations which do not keep cows.

Evolution by cultural means has been construed in a very different way by Richard Dawkins, not as a new *mechanism* of evolution, but as the creation and replication of a new kind of entity – the *meme*. A meme is a scrap of culture – a slogan, a song, an idea. Memes inhabit brains, and replicate themselves by being transmitted from one brain to another, somewhat analogously to the way genes inhabit bodies.

It is not surprising that Dawkins should construe evolution by cultural means in these terms. His gene-centred, purpose-depleted vision of evolution obstructs thinking of evolution by cultural means as the development of a new, quasi-Lamarckian method for the reproduction of purposive ways of life, grafted onto genetic reproduction. Meme replication and evolution seem to mimic gene replication and evolution; it is understandable, therefore, that Dawkins should want to construe evolution by cultural means in these terms.

How do the two versions of evolution by cultural means compare? Darwin(5) is broader in scope, in that new actions that have evolved by cultural means, such as the chimpanzee trick for getting termites to eat, need not constitute memes. Darwin(5), because of its emphasis on the evolution of *purposive action*, brings to the fore, and renders explicable, the way in which evolution by cultural means can have an impact on subsequent evolution of bodily changes determined genetically, in a way in which Dawkins' meme view cannot. In other words, Darwin(5) brings to the fore the fact that non-genetic evolution of behaviour can help bring about subsequent evolution brought about by genetic changes. Finally, and following on from the last point, Darwin(5) very strikingly reveals how elements of purposiveness can be incorporated into the *mechanisms of evolution* themselves. The view helps us understand how Darwinian evolution of animal life can, seamlessly, become purposive human history. The meme view does not do this.

Do those concerned with evolution – biologists, anthropologists, archaeologists, psychologists and others – appreciate just how fundamentally evolution by cultural means transforms the orthodox conception of Darwinian evolution? I am not at all sure that they do.

For the first six or seven decades of the 20th century, social scientists treated cultural or social evolution of humans as if this were quite distinct from Darwinian evolution.[10] But then, associated with an explosion of interest in Darwinian theory, social scientists began to appreciate that Darwinism has far-reaching implications for the social sciences, and for social or cultural evolution in particular. Around 1980, a number of evolutionary thinkers realized that cultural evolution cannot be dissociated from Darwinian evolution because genetic and cultural evolution interact with one another. Works began to appear that recognized this interplay of genetic and cultural evolution, or *coevolution*. There is, for example, *Culture and the Evolutionary Process* by Robert Boyd and Peter Richerson (1980), and their subsequent *Not By Genes Alone* (2005); there is *Human Culture: A Moment in Evolution* by Theodosius Dobzhansky[11] and Ernest Boesiger (1983); there is *Coevolution: Genes, Culture, and Human Diversity* by William Durham (1991); and there is *The Evolution of Culture*

---

[10] Donald Campbell put the matter like this. Having referred to a body of work by social scientists on social evolution published between 1950 and 1970, he remarks "In all of this, social evolution is seen as a separate process from biological evolution, although made possible by it" (Campbell, 1975, p. 1104-5).

[11] In this book the interaction of cultural and gene-based evolution is at least acknowledged, although earlier, as we shall see, Dobzhansky dismissed the so-called "Baldwin effect": see Dobzhansky (1970, p. 211).

edited by Robin Dunbar *et al.* (1999). Then there are a number of works that propound some specific theory concerning some aspect of human evolution, but in a way which presupposes gene-culture coevolution, such as *The Scars of Evolution* by Elaine Morgan (1990), *Grooming, Gossip and the Evolution of Language* by Robin Dunbar (1996), and *The Mating Mind* by Geoffrey Miller (2001). Social scientists, as one might expect, tend to concentrate on evolution by cultural means as it affects human evolution. Others have, however, studied evolution by cultural means in animals: see, for example, *The Evolution of Culture in Animals* by John Bonner (1980). A book that covers both is *Social Evolution* by Robert Trivers (1985).

In view of the extensive literature on gene-culture coevolution (of which the above is but a glimpse), what grounds can I possibly have for declaring that evolutionary biologists do not sufficiently emphasize the fundamental role of *purposive action*, of behaviour, learning and culture in evolution? What I find lacking is an awareness of just how widespread and fundamental is the role that goal-pursuing action plays in evolution, and how dramatically the theory needs to be reformulated to take this role fully into account. It means that the theory needs to be reformulated to take into account that *the mechanisms of evolution themselves evolve* as purposive action, learning, imitation, culture and evolution by cultural means come to play increasingly significant roles.

Darwinians are scornful of the idea that the actions of the giraffe's ancestors, in stretching to eat leaves high up in trees, had any causal role in producing the present-day giraffe's long neck. This, it is claimed, is utterly discredited Lamarckism. Thus do Darwinians reveal their failure to appreciate just how fundamental is the role of purposive action in evolution. For, of course, Lamarck is right – or partly right. *The stretching of the neck of the giraffe's ancestors does play a vital causal role in the subsequent development of the giraffe's long neck.* Stretching does not directly cause offspring to have longer necks. There is here no inheritance of acquired characteristics. But if ancestors had not stretched their necks, the modern giraffe would not possess its long neck. Ancestors stretching their necks to reach leaves good to eat is not *sufficient* for offspring to develop long necks. But it is *necessary*. Only then do mutations that lead to longer necks have survival value, and thus spread through the population. It is reasonable to hold that, throughout the animal kingdom, purposive action leads the way. Beaks, teeth, tusks, camouflage, claws, muscles, horns, hooves, digestive systems, and other bodily characteristics only develop because animals are living in a certain way in a certain environment and, relative to these, the bodily characteristics in question have value from the standpoint of survival and reproductive success. As I have said life unknowingly *breeds itself into existence*.

Thus, how an animal lives crucially affects what has survival value, and this in turn crucially affects the animal's evolution. In particular, changes in the way a kind of animal lives, which may come about because of genetic changes, environmental changes, or evolution by cultural means, can have dramatic consequences for that animal's subsequent evolution. It may well be that changes in ways of life play the leading role in evolution.[12]

I remember well the way in which my whole perception of evolution changed dramatically as a result of becoming aware of the all-pervasive influence of purposive action on evolution, some time in the late 1960s. This came about as a result of three events. First, there was a stray remark of J. Z. Young during a lecture at University College London on the brain. He remarked that the way an animal's memory worked would depend on how it lived. With a shock I realized the obvious: evolution designs brains, and

---

[12] To the demand that evidence is required to substantiate this thesis, my response would be that the thesis is an all-but straightforward implication of Darwinism. Changes in ways of life are bound to change what has value for survival and reproductive success, and this in turn over time is bound to have consequences for gene-based evolution in a majority of cases. Furthermore, changes in ways of life are bound to occur, as a result of environmental changes, changes of habitat, changes in predators, food supply, or competitors, or changes brought about by learning. The boot is on the other foot. What requires establishing is that changes in ways of life only rarely affects subsequent gene-based evolution. In the absence of evidence for this thesis, we should adopt the Darwinian view that changes in purposive action widely, even generally, lead the way in subsequent gene-based evolution.

therefore minds. And how an animal lives affects how it evolves. The second event was a remark of Karl Popper during a lecture at the London School of Economics. He made the point that fish acquired the capacity to emerge from the sea and live on land only because certain fish took to living in shallow water near beaches, thus becoming stranded in pools every now and again as the tide retreated. Living in this way, developing the capacity to breathe air, and move across land, would have had great survival value, something which would not have been the case for fish living in the deep ocean. How the animal lives, in short, crucially affects its subsequent evolution. Purposive action, and *changes* in purposive action, may well be at the leading edge of animal evolution quite generally. The third event was reading Alister Hardy's *The Living Stream* (1965), of which more in a moment.

The outcome was a profound shift in the whole way in which, it seemed to me – and still seems – evolution needs to be understood. The actions of animals, our ancestors, in the past, for millions of years, have had a vital role to play in bringing about our existence. Evolution is not just blind chance and necessity, to quote the title of a book by Jacque Monod (1974). Our animal ancestors, striving to live, to eat, to avoid being eaten, to mate, to rear young, are a vital part of the reason for our existence. They did not, of course, intend us to exist. Nevertheless, without their striving, we would not be here. We really do owe them a debt of gratitude.[13]

Major development in Darwinian theory took place around 1930 with the rediscovery of the work of Mendel on genetics, and its incorporation into the theory of evolution, and associated with the work of R. A. Fisher (1930), J. B. S. Haldane (1932) and Julian Huxley (1942). We ought to recognize that a similarly dramatic development in Darwinian theory took place some time in the 1980s with the incorporation of elements of Lamarckism into the theory. Lamarck was wrong to hold that acquired physical characteristics are inherited. He was right, however, in his view that purposive action plays a vital role in subsequent evolution of physical characteristics. And he was right to hold that acquired

---

[13] For my earlier accounts of this purposive version of Darwinism see the references given in note 6.

characteristics are inherited: this occurs in evolution by cultural means, the acquired characteristics being learned purposive actions that are passed on by imitation. Lamarck did not get everything right, but who does? Even Darwin made mistakes.

Why has the vital role that purposive action plays in Darwinian evolution not received the proper emphasis it deserves in modern accounts of the theory? A part of the reason may be that discredited Lamarckism has formed an intellectual barrier in the minds of evolutionists to recognizing the Lamarckian character of Darwinian evolution.[14] Another, possibly related reason, has to do with a failure of evolutionary biologists to get the history of the idea right. The idea that evolution by cultural means can have an impact on subsequent gene-based evolution is usually attributed to Mark Baldwin, and is usually known as "the Baldwin effect". Baldwin did indeed publish a version of the idea long ago in 1896.[15] G. G. Simpson appears to have introduced the phrase "The Baldwin Effect" (Simpson, 1953). It is extensively discussed in Daniel Dennett's *Darwin's Dangerous Idea*.[16] There is even an entire book devoted to the subject, the outcome of a conference, with the title *Evolution and Learning: The Baldwin Effect Reconsidered* (Weber and Depew, 2003). But the idea did not come, originally, from Baldwin. And in so far as Baldwin expresses the idea, he does so badly. Some of the modern accounts of the idea are even worse.[17] As expressed by Simpson, and by many since who have followed him, "the Baldwin effect" amounts

---

[14] Orthodox Darwinians do, it is true, acknowledge the Lamarckian, or "quasi-Lamarckian", character of evolution by cultural means: see, for example, Cronin (1991, p. 373). Boyd and Recherson (1985) are prepared to say that cultural evolution creates "a kind of 'Lamarckian' effect" (p. 9). But the Lamarckian character of Darwinian evolution *per se* is not acknowledged. On the contrary, it is fiercely resisted. This further substantiates my point that orthodox Darwinians do not fully recognize and acknowledge the vital and general role that purposive action plays in evolution.

[15] See Baldwin (1896a; 1986b).

[16] Dennett (1996), pp. 77-80, 190, 300, 322-3, 338, 374, 403n, 463.

[17] Dennett's characterization of "the Baldwin effect" is peculiarly opaque for an author usually so lucid: for the reference, see previous note.

to this. A new kind of action in a group of animals that comes about as a result of learning, is eventually determined genetically. But this falls to the obvious objection: if an animal has learnt to act in a certain way, what possible value for survival and reproductive success can there be in having this learnt action become such that it is determined genetically? There would be no selective pressure for this to occur. And "the Baldwin effect" has been dismissed on just these grounds, by Simpson (1952), by Mayr (1963, pp. 610-2), by Dobzhansky (1970, p. 211), and others. Depew, for example, puts the objection like this: "If learned behaviors are so effective in getting a useful trait passed from generation to generation at the cultural level, there will presumably be no selection pressure for the spread of genetic factors favoring that trait" (Weber and Depew, 2003, p. 15).

But all this represents a catalogue of errors. Evolution by cultural means was first put forward independently by Lloyd Morgan (1896) and Fairfield Osborn (1896). Baldwin took the idea from Lloyd Morgan, subsequently took the credit for it, and then failed to do the idea justice.[18] Lloyd Morgan's idea, of course, is not that a new kind of purposive action, passed on by imitation, eventually becomes determined genetically, but rather that new purposive actions generate new selective pressures, and mutations which create traits which *facilitate* the new actions will be selected for. Thus, if a dog-like creature takes to catching fish in rivers, it is not this new action which will become genetically determined: rather, mutations which tend to transform legs into flippers will be selected for, given the new way of life. Evolution by cultural means has not been given the importance it deserves in part because it has so often been understood in a peculiarly bungled form, which renders the idea untenable. Finally, as we have seen, it is reasonable to hold that purposive action plays a vital and widespread role in evolution even when evolution by cultural means is not involved. Cultural evolution may not have been involved in the giraffe acquiring its long neck, but purposive action was involved.[19]

---

[18] See Hardy (1965), pp. 164-9 and Bateson (2004), pp. 286-9 and 290-1.
[19] Darwin(4) incorporates purposive action into the mechanisms of

There is one book that does do justice to the idea that purposive action is at the leading edge of Darwinian evolution – and to the history of the idea: Alister Hardy's *The Living Stream* (1965). Hardy sums up the idea as follows:

> If a population of animals should change their habits (no doubt, often on account of changes in their surroundings such as food supply, breeding sites, etc., but also sometimes due to their exploratory curiosity discovering new ways of life, such as new sources of food or new methods of exploitation) then, sooner or later, variations in the gene complex will turn up in the population to produce small alternations in the animals structure which will make them more efficient in relation to their new behaviour pattern; these more efficient individuals will tend to survive rather than the less efficient, and so the composition of the population will gradually change. This evolutionary change is one caused *initially* by a change in behaviour (Hardy, 1965, p. 170).

Hardy begins with the evolution of camouflage, the effectiveness and cleverness of which is the outcome of perceptive predators seeing through early ineffective efforts – a clear case of unknowing breeding. He then goes on to expound and illustrate the way in which Lamarckian evolution by cultural means has consequences for subsequent gene-based evolution. And he discusses the history of the idea: the contributions of Lloyd Morgan, Osborn and Baldwin, and the subsequent contributions, of one kind or another, by Simpson (1953), Huxley (1942), C. H. Waddington (1957) and others. He quotes E. S. Russell (1916) as declaring "We need to look at living things with new eyes and a truer sympathy. We shall then see them as active, living, passionate beings like ourselves, and we shall seek in our morphology[20] to interpret as may be their form in terms of their activity" (Hardy, 1965, p. 181). He quotes E. Schrödinger (1958) as asserting that "Without changing

---

evolution but does not involve evolution by cultural means.
[20] The study of the form of living things.

anything in the basic assumptions of Darwinism we can see that the behaviour of the individual , the way it makes use of its innate faculties, plays a relevant part, nay, plays the most relevant part in evolution" (Hardy, 1965, p. 189).[21] Hardy also quotes a passage from James Hutton which beautifully expresses a version of the idea, written "a hundred years before Lloyd Morgan and Baldwin put forward their versions of the theory...eleven years before Charles Darwin was born and twelve years before Lamarck first published his evolutionary views" (Hardy, 1965, p. 179).

Given all this, one might suppose that Alister Hardy is, today, hailed as a major figure in launching the idea that purposive action plays a key role in evolution – the key idea of what I have called Darwin(4) and Darwin(5). Not at all. He is rarely mentioned. Boyd and Richerson (1980) and (2005); Durham (1991); Dunbar (1996) Dunbar *et al* (1999), Miller (2001), Bonner (1980), and Trivers (1985) make no mention of Hardy whatsoever. Dennett (1996) does refer to Hardy, but only as the author of the aquatic ape hypothesis, and in connection with Elaine Morgan's long-standing and brilliant championing and development of the idea. Even Morgan does not refer to Hardy's *The Living Stream*, although much of her work illustrates the key idea of that book.[22] Peter Bowler's *Evolution: the History of an Idea* (2009) does mention Hardy, but only to say "Hardy (1965) openly endorsed the Baldwin effect" (p. 367). Furthermore, this is in the context of discussing Arthur Koestler's anti-Darwinian ideas.[23] Of the 14 contributors to *Evolution and Learning: The Baldwin Effect Reconsidered* (Weber and Depew, 2003), only one mentions

---

[21] Unfortunately, as Hardy points out, Schrödinger goes on to retract much of the content of this splendid brief statement of the basic idea.

[22] The aquatic ape hypothesis holds that pre-human ancestors lived on the shores of rivers, lakes and the sea, and spent time in the water, many of our bodily characteristics stemming from that way of life. This illustrates the general idea that a way of life has consequences for gene-based evolution.

[23] But to be fair to Bowler, he gives a good brief formulation of the misnamed Baldwin effect: "new habits are supposed to determine which genetic variations are most useful" (Bowler, 2009, p. 367). The word "supposed" here does not, however, exactly inspire confidence.

Hardy, and only very briefly and obliquely, in connection with a letter of Waddington to Hardy (p. 146). There is, however, a critical essay review of the book by Patrick Bateson (2004) which emphasizes the importance of Hardy's contribution, and refers to the work of others along similar lines as well.

Alister Hardy's own explanation for the neglect of his thesis and book was that, in the penultimate chapter he went on to defend telepathy. This may well have played a role. In any case, the continuing neglect of Hardy's work is symptomatic, I claim, of a continuing failure, on the part of evolutionary biologists and social scientists, to do justice to the profound transformation in Darwinism that is brought about when one acknowledges the vital and general role that purposive action plays in evolution. Only then does one recognize that the mechanisms of evolution themselves evolve, as they assign increasingly important roles to purposive action, and become Lamarckian in character, as we have seen.

So far nothing has been said about sentience or consciousness. I now repair that omission with the next version of Darwinism.

Darwin(6). Sentience. Purposive explanation becomes increase-ingly important as we move from Darwin(3) to Darwin(5). At a certain time (or perhaps independently at a number of different times) living creatures became *sentient*, and what may be called *sentient* explanation became relevant to evolution, in Darwin(3) to Darwin(5) ways. (Sentient explanations take the inner sensations of the creatures in question into account. All sentient explanations are purposive, but not *vice versa*.)

We saw in chapter 3 that the experiential cannot be derived from science that is reducible (in principle) to physics. We can take this to imply that sentient explanations cannot be derived from purposive ones. This means Darwinian evolution cannot, even in principle, explain how the sentient has emerged or evolved from the non-sentient, purposive, neurological and physical. Necessary and sufficient neurological conditions for sentience do however, presumably, exist, and we can speculate as to what these may be. Elsewhere I have suggested that sentience emerges with the transition from what may be called *sequential* to *motivational* control.[24]

There is a wasp that lays its egg and then flies around, clutching the egg, looking for a suitable hole in the ground in which to bury it. When it finds a candidate, it puts the egg down at a fixed distance from the hole, goes into it to see if it is suitable and, if it is, comes out, fetches the egg and buries it. All this looks as if the wasp knows what it is doing. However, if the egg is moved a bit further away while the wasp is investigating the hole, the wasp will emerge, pick up the egg, place it at the fixed distance from the hole, and then investigate the hole again for suitability. This can be repeated many times. Evidently, the wasp is led by its brain to do one specific kind of action (fly around looking for a hole, or put the egg down at a fixed distance from the hole, etc.) until, it is triggered to move onto the next specific action in the sequence of actions. The wasp has no idea of its overall goal (to bury the egg in the ground). It achieves this goal by achieving a sequence of precisely specified intermediate goals, the completion of one triggering pursuit of the next in the sequence. This is what I mean by "sequential control".

It is very different, I surmise, from the way tigers are controlled by their brains to go hunting. The overall goal – eating food – is actively represented in the tiger's brain, and the brain has to work out what actions have to be performed if the goal is to be realized. This means the tiger may, on different occasions, perform different actions in order to attain the final goal of eating. This is what I mean by "motivational control".

I conjecture that it is the transition from sequential to motivational control which leads to the emergence of sentience. The goal of eating, represented in the brain by means of characteristic neurological activity *is*, for the tiger, the *feeling* of hunger, the *desire for food*.

From the standpoint of survival value, motivational control has the great advantage over sequential control that it is flexible. It is open to leading to a variety of different actions in different circumstances, in ways in which the rigid sequence of specific

---

[24] See Maxwell (2001), ch. 7, and especially pp. 180-185.

actions of sequential control is not. Motivational control makes *learning* possible, in ways in which sequential control does not. On this view, then, feeling and desire are at the core of sentience, and sentience evolves so that actions of animals can be specified in an open-ended, flexible way, allowing for learning, and for actions to be adapted to circumstances.

It is often remarked that what matters, from a Darwinian perspective, is what you *do*, how you *act*, not what your inner *feelings* and *desires* are.   Darwinian evolution thus seems peculiarly ill-equipped to help explain how and why sentience and consciousness have come into existence.   From a Darwinian perspective, acting *as if* sentient or conscious is just as good as actually *being* sentient or conscious (assuming that sentience and consciousness confer some selective advantage).   From a Darwinian perspective, it seems, we might as well all have been zombies.

This echoes the problem we encountered in the last chapter, in connection with free will. The solution put forward there must be employed here. In our world, zombies are not possible. Brains that perform sufficiently sophisticated motivational control are automatically sentient.

It may be asked: Which comes first, evolution by cultural means or sentience? Does learning and imitation, of the kind required for evolution by cultural means, presuppose sentience? My guess is that it does (although it would seem to me possible to design non-sentient robots able to participate in evolution by cultural means).

Darwin(7).   From sentience there emerges consciousness. Conscious action begins to play a role in evolution, in ways specified by Darwin(3) to Darwin(5) for purposive action. That is, conscious action replaces unconscious - even insentient - purposive action.   Conscious beings choose mates, rear offspring, make discoveries and imitate the discoveries of others, aware of what they are doing.

Comments.   What factors are behind sentience evolving into consciousness, as we human beings know it? I suggest three: (a) imagination, (b) personalistic understanding of others (and so of oneself), and (c) language. Let us take these in turn.

291

(a) To imagine you are climbing a mountain (when you are not doing anything of the kind) is to give yourself experiences somewhat like what you would have were you actually to be climbing a mountain. It is, in other words, to make occur in your brain neurological processes somewhat like those that would occur if you were actually to be climbing a mountain. Being able to imagine clearly has potential survival value. It means you can try things out in the imagination, thus learning from imaginative failure and success, far less risky, far less time and energy consuming, than trying things out in reality. Better to die in the imagination, than die in reality. I have suggested that it may be the function of dreaming to develop the capacity to imagine in the individual who dreams. Imagination makes possible a vast increase in the arena of action. As a result of imagining one is at other places, other times, it becomes possible to become aware of distant places and times – both the distant past and future. The discovery of the inevitability death becomes possible – something likely to have had a big impact on human evolution and history. It may well be that it is the development of the capacity to imagine that is the crucial step from sentience towards human consciousness. For it is this that makes inner action possible – doing things in the imagination. This would seem to be the crucial distinction between consciousness and sentience – whether one can do things internally, in imagination – act, explore, think, ponder, question – or whether one is condemned merely to feel, to desire, to experience sensations. We can perhaps see, however, how imagination could develop from motivational control. To have actively present in your mind the goal you seek to achieve – food, a mate – is already close to imagining you are doing what you are not: eating freshly killed prey, mating. On this view, motivational control first spawns sentience, then imagination, then consciousness.

(b) Being able to imagine you are doing things that you are not doing makes it possible to imagine that you are a person that you are not. It makes personalistic understanding possible, in other words.[25] Being good at understanding others can clearly have

---

[25] One might think that personalistic understanding of others is a

survival value when animals or humans are living socially. It is needed to divine the intentions of others, to form alliances and friendships, and perhaps to mate and reproduce. As a result of understanding others, one understand that others understand one's self, which in turn can create an awareness of discrepancies between how others see one, and one's own experiences of one's self. It is this awareness of this discrepancy, I suggest, which creates self-consciousness. As a result of becoming aware of others' awareness of oneself, one becomes are of what those others are not aware of and do not experience: one's own experiences. Thus one acquires self-consciousness.

(c) It has been shown by Paul Grice[26] that human communication involves, quite essentially, multi-layers of mutually understood intentions. If I am to communicate with you by means of language, I must intend this, you must understand that I intend it, and I must understand that you understand. These multi-layers of implicitly understood intentions will have evolved gradually from their beginnings in primate, one layered animal communication. Let us suppose A communicates to B. Human communication, I conjecture, has evolved by means of the following steps:

(i) A acts in its own interests, for example goes rapidly into flight to avoid a predator; B takes this behaviour as an indication of something (in this case danger), for him, and acts accordingly.

---

straightforward specific use of imagination. It may be, however, that understanding others develops separately from the development of the ability to imagine, and uses different parts of the brain. The discovery of so-called "mirror" neurons, used to understand others "empathically" or personalistically, would seem to suggest that this is the case – although this role of mirror neurons has been called into question. That being able to imagine is distinct neurologically from being able to understand others is further suggested by autistic people, who may have vivid imaginations, and yet be poor at understanding others (imagining they are other people).

26 Grice (1957), reprinted in Grice (1989), ch. 14. Grice makes no attempt, however, to indicate, as I have done here, the manner in which the multi-layered character of human communication can be seen as having emerged gradually as a result of Darwinian evolution.

(ii) In addition, A does something which is such that the sole purpose of it is to communicate to B, even though A has no such conscious intention. Here A might squawk as it goes into flight in a manner characteristic for that species in such circumstances; B reacts accordingly.

(iii) In addition, A has the purpose of signalling to B since, if A knows that it is on its own it will not signal (e.g. squawk).

(iv) In addition, A has the purpose of communicating the message of the action to B, so that, in the case of the squawk, the bird squawks in order to warn B. If B is present but in no danger then the bird does not squawk.

(v) B understands the message, the meaning of the squawk.

(vi) A has the purpose of B understanding the meaning of the message.

(vi) B understands this too.

(vii) A intends B to understand this.

And so on (the multi-layers of mutual understanding, initially profoundly significant, as one goes on further, becoming increasingly insignificant).

Communication and language evolve in tandem with the evolution of consciousness and personalistic understanding (or empathy), each requiring, but also enriching, the other two. Communication up to (iii), or perhaps (iv), does not require conscious intention and personalistic understanding, but from (iv) or perhaps (v) onwards, the kind of communication involved does require consciousness and personalistic understanding.

One profoundly significant consequence of imagination is that it enables the imagining creature vastly to increase the arena of its actions. As a result of imagining one is doing what one is not doing, it becomes possible to imagine one is doing things at other places and times: one can become aware of distant places, the distant past and future. The discovery of the inevitability of death becomes possible. Long term conscious planning becomes possible. All this will be further strengthened and enriched by the associated development of personalistic understanding, language and communication and, with these, the development of a common, shared public world. These developments make possible:

294

Darwin(8). Ways of life evolve as the outcome, in part, of conscious actions. Darwinian evolution gives rise to history.

Comments. In order to understand how and why human affairs evolve as they do, one must take note of the conscious intentions and actions of people. This is not to say that what happens is invariably the outcome of what people intend. People's intentions and plans conflict: there are winners and losers. Often there are no winners, and the outcome is intended by no one. And quite apart from conscious intentions and actions, natural phenomena also play their part: storms, droughts, plagues, earthquakes. There are also the unintended consequences of human actions to be taken into account as well: traffic congestion, depletion of natural resources, extinction of species and environmental degradation, climate change.

There is a tendency to think that history is associated exclusively with human beings. There are grounds for holding, however, that this is not quite correct. Even chimpanzees produce history, to a limited extent. Groups of chimpanzees engage in power struggles, alliances, long-standing disputes, and even war.

Among the dramatic consequences that the emergence of consciousness has for the evolution of ways of life, two that deserve to be highlighted are (1) the discovery of the inevitability of death, and (2) conflict over control of way of life. Let me take these in turn.

(1) The discovery of the inevitability of death. It seems likely that most mammals, being confined to the here and now, do not realize that their eventual death is inevitable. Once our imagination has roamed far and wide, however, we can foresee a time in the distant future when we will grow old and die. Imagination has expanded the arena in which we act and live sufficiently to encompass our future death. Granted that evolution assigns to us survival and reproductive success as our fundamental goals in life, all other gaols being means to these ends, the discovery of our inevitable death must be traumatic indeed. Imagination, which made the discovery, quickly steps in and seeks to find ways to evade the grim news. It is our imagination that creates the possibility that death is but the beginning of a journey to another, and possibly happier, place. Anthropologists take

burial remains as a good indication of early human culture. Often, burial remains include pots of food to help the traveller on his or her way to life after death. As history has unfolded, people have found other ways to try to cheat death, besides having children: memorials, pyramids, enduring art, contributions to science, thought and literature, fame, conquest, institutions and corporations. Death remains a potent problem in human life. Discovering how to cope with its inevitability is no doubt important if one is to achieve a happy, mature way of life.

(2) It is reasonable to hold that most mammals do not consciously plan their way of life. They consciously[27] plan what they do from minute to minute, but the way of life is planned by an unconscious master control system, which directs consciousness to seek appropriate goals – food, shelter, mating, fighting, rearing of young – by planting into consciousness relevant desires and feelings of hunger, fear, desire, by means of hormones, sugar content in the blood, etc. The master control system pulls the strings, and consciousness leaps to obey. Consciousness is the slave of the master control system, and remains so as long as imagination is not sufficiently developed to enable consciousness to become aware of what lies beyond the immediate here and now. But the moment imagination, helped out by personalistic understanding and language, becomes aware of the distant past and future, and other places, it becomes possible for consciousness to attempt something evolution has not equipped it to do: take charge of, and plan, the way of life. As long as people lived in hunting and gathering tribes, living from day to day more or less as

---

[27] I am employing "consciously plan" somewhat ambiguously here. I assume sentient animals may, in some sense, be said to "consciously plan" actions, from moment to moment, even if these animals are not fully self-conscious in the way we human beings are. I also assume that animal moment-to-moment "conscious acting" is a precursor of our "conscious acting", ours having evolved from the more primitive version to be found in animals. This is, perhaps, the key to consciousness. At its most primitive, it is what controls action, in ordinary circumstances, from moment to moment, in sentient animals. It is this that is enriched with the evolution of (a) imagination, (b) personalistic understanding and (c) language, to become human self-consciousness.

chimpanzees do, the potential conflict between the master control system and consciousness would not have been too pronounced. It is when humanity departed from the hunting and gathering way of life by taking to agriculture, which requires conscious action directed towards long-term goals (planting, weeding, growing and harvesting of crops), that the conflict would have become active, consciousness grappling with and seeking to over-ride the master control system, and all too often failing in the attempt. This conflict lies at the heart of our existence today. The master control system is especially good at tricking consciousness into thinking it is in charge when actually it is the master system which controls the action. Rationalization, in other words, is rife in human affairs. Distorted versions of the conflict emerge in religions, and in psychoanalytic theory. We have still failed to get it properly into perspective. What is required, of course, is that we acknowledge the existence and nature of the conflict, and find the best way to resolve the conflicts that will inevitably arise as we live.[28]

Darwin(9). History is under conscious control in pursuit of life of greater value, in so far as such a thing is possible.
Comments. For this to come about, it is essential that humanity adopts and implements wisdom-inquiry and its methodology, aim-oriented rationality. Seen in this light, wisdom-inquiry and aim-oriented rationality complement and complete Darwinian theory. Darwinian theory, properly understood, brings graphically to the fore that the fundamental task that lies before humanity is (1) to get conscious control of history, and (2) to discover how life-aims, bequeathed to us by evolution and history, can be transformed into the aim of discovering and realizing what is of value in life. Crudely put, our fundamental problem is to transform the Darwinian life-aims of survival and reproductive success into the life-aims of survival, reproduction and enrichment of *life of value*. Our evolutionary and historical past has ill-equipped us for this task. Evolution manufactures species in great abundance able to adopt a great variety of means to realize a *fixed* aim: survival and

---

[28] My comments on Darwin (4) to Darwin(8) are based on, and develop, earlier remarks of mine on evolution: see works referred to in note 6.

reproduction. Nothing in evolution has equipped us to *transform* our basic life aim. Our culture, inherited from the past, is not designed to help. Far from helping us progressively *improve* our life aims towards promoting *life of value*, all too often it disguises from us the real nature of our aims. This is true even of one of our most successful endeavours, science – as we saw in chapter 5. As yet, we have not even appreciated the nature of our problem: to transform our basic life aims – personal, institutional, global. We have failed to grasp what we require to help solve this fundamental problem, namely wisdom-inquiry, and aim-oriented rationality. Even philosophers ignore the fundamental problem: *What kind of inquiry can best help us make progress towards as good a world as possible?. What kind of inquiry can best help the God-of-Cosmic-Value to flourish within the God-of-Cosmic-Power?*[29]

During the time that it has taken me to write this book, three global issues have demonstrated graphically just how disastrous is our failure to put aim-oriented rationality and wisdom-inquiry into practice – our failure, indeed, even to have the idea that this is what we need to do!

The first is George Bush's disastrous "war against terrorism", a war fought in such a way that it has all-but transformed the USA into a terrorist nation itself, and has acted as a magnificent

---

[29] In other words, Darwinian theory, properly appreciated, powerfully endorses the point, at present almost universally ignored, that it is of fundamental importance for us to build wisdom-inquiry and aim-oriented rationality into our social world and culture, if we are to enhance our capacity to realize what is of value in life. But this point is also powerfully endorsed by the compatibilist view of free will I argued for in chapter 7. According to that compatibilist view, any free will we possess in excess of that possessed by chimpanzees, let us say, is due in large part to our language and culture, our technology and social institutions. Brought up without these, like Truffaut's wild child, our free will – our capacity to realize what is of value – would not amount to much more than that of a chimpanzee. In order to enhance our free will – and enhance free will in the sense of the capacity to realize what is of value in life – it is vital that our social world and culture is designed to facilitate their flourishing. Darwinian theory, wisdom-inquiry and compatibilism form a kind of interlocked trinity, each component powerfully reinforcing the other two.

recruiting agent for terrorists. The failure even to consider that the aims being pursued might have almost the opposite consequences to genuinely desirable ends could hardly be more obvious.[30]

The second is the global financial crisis of 2008. It should have been obvious, some years before that date, that international banking was being conducted in an unsustainable way. All that would be required to bring the financial system crashing down was a fall in property values, a crisis of confidence. A few voices did indeed cry out that we were heading towards disaster, but they were ignored. There was the most elementary failure to consider the likely dire consequences of pursuing the then current aims of banks a mere ten years into the future. Nothing could demonstrate more graphically our failure to understand the need for, let alone implement, basic ingredients of aim-oriented rationality.

The third global issue I have in mind is global warming. This again illustrates the profoundly problematic character of basic aims. At one time it seemed that the aim of developing modern industry, transport and agriculture on a world-wide basis could only be good. Unfortunately, among other undesirable consequences, it leads to an increase in carbon dioxide in the atmosphere which, in turn, leads to a rise in average global temperatures, the melting of glaciers and polar icecaps and the rising of sea levels. If we continue as we are, we face catastrophe. What once seemed so desirable now threatens our existence. We have known about impending global warming since 1960, if not earlier, but it has taken a very long time for those with the power to determine what we do – politicians, industrialists, entrepreneurs – to begin to consider what needs to be done to cut CO_2 emissions down. A perceived threat from a neighbouring country could galvanize a nation to prepare for war in a matter of months. Global warming, which threatens the very future of humanity is, it seems, too impersonal or unfamiliar to provoke a similar response. So far we have failed to change our lives in the ways required to avoid disaster. If wisdom-inquiry and aim-oriented rationality had been in place since 1960, academia would have been shouting from the rooftops for the need to change. Our world would long ago have

---

[30] See, for example, my (2007b).

been alerted to the urgent need to change its ways. Without them, it has taken decades for news of the seriousness of our situation to filter through to those able to take action.

Put our human world into the context of Darwinian evolution and what cries out is the urgent need to put into place, in our human world, strategies and modes of thought designed to help us improve our problematic aims, personal, institutional, global, in the direction of promoting long-term life of value, as we live. First of all, Darwinian evolution selects for the capacity to survive and reproduce in the short term. As Steve Jones has wittily put it, evolution has tactics but no strategy. It does not bequeath to us the life aim of *long-term life of value*. This is an aim we must painfully acquire through modifying progressively what we have inherited from evolution and history. Secondly, Darwinian evolution leaves us peculiarly ill-equipped to transform our life aims in the way we require for, throughout evolution, there is one fixed fundamental aim: survival and reproduction. The basic Darwinian lesson is: *we both must, and are peculiarly ill-equipped to, transform our basic life aims.* How striking it is that the volumes of print produced on the social implications of Darwin have so rarely come up with this simple, stark, vital Darwinian lesson.

There is also, however, a more hopeful message that emerges from a consideration of our human world in the context of Darwinian evolution. What distinguishes us from all other species is the massive extent to which we are the product of evolution by cultural means. We have this unique capacity to learn. There is hope. We may be able to learn how to improve our problematic – even destructive – aims as we live, even though evolution ill-equips us for this task.

In the next, and final chapter, I shall spell out in a bit more detail what in my view we need to do if we are to do better at making progress towards as good a world as possible, thus helping the God-of-Cosmic-Value to flourish less painfully and somewhat more luxuriantly within the cold, remorseless embrace of the God-of-Cosmic-Power.

CHAPTER NINE

OUR GLOBAL PROBLEMS AND WHAT WE NEED TO DO ABOUT THEM

Global Problems
The Role of Modern Science and Technology
What Do We Need to Do?
How Could Wisdom-Inquiry Help?
Objections
Is the Academic Revolution Underway?
Conclusion

Global Problems

Can humanity help the God-of-Cosmic-Value to flourish, ensnarled as it is within the remorseless grip of the Cosmic-God-of-Power? Can we, in other words, successfully realize what is genuinely of value to us in the real world – more successfully, at least, than we have managed to do so far, up to the first decade of the 21st century? Much depends, I will argue in this final chapter, on whether we succeed in putting wisdom-inquiry and aim-oriented rationality into practice in academia, and in life.

As I have already stressed, we are confronted by grave global problems. There is the problem of vast differences in wealth around the globe, something like a third of the world's population living in conditions of dire poverty, without enough to eat, safe water, proper shelter, health care, education, employment. Over 9 million children die every year from preventable causes – some 25,000 every day.[1] There is the problem of war, over 100 million people having died in wars during the 20th century, which compares unfavourably with the 12 million or so who died in wars in the 19th century. And we have not been doing very well in the first decade of the 21st century (I am writing in July 2009). There is

---

[1] See www.unicef.org/media/media_45485.html.

the problem of the spread and stockpiling of deadly modern armaments, even in poor countries, and the ever-present threat of their use by terrorists or in war, whether the arms be conventional, chemical, biological or nuclear. Nuclear proliferation is an especially grave problem, India, Pakistan and north Korea having recently acquired the bomb, and other nations, such as Iran, likely to acquire the bomb soon. There is the long-standing problem of the rapid growth of the world's population, especially pronounced in the poorest parts of the world, adversely affecting efforts at development. There is the problem of the progressive destruction of tropical rain forests and other natural habitats, with its concomitant devastating extinction of species. And there is the horror of the aids epidemic, again far more terrible in the poorest parts of the world, devastating millions of lives, destroying families, and crippling economies.

And, in addition to these stark global crises, there are problems of a more diffuse, intangible character, signs of a general cultural or spiritual malaise. There is the phenomenon of political apathy: the problems of humanity seem so immense, so remorseless, so utterly beyond human control, and each one of us, a mere individual, seems wholly impotent before the juggernaut of history. The new global economy can seem like a monster out of control, with human beings having to adapt their lives to its demands, rather than gaining support from it. There is the phenomenon of the trivialization of culture, as a result, perhaps, of technological innovation such as TV and the internet. Once, people created and participated in their own live music, theatre, art, poetry. Now this is pumped into our homes and into our ears by our technology, a mass-produced culture for mass consumption; we have become passive consumers, and the product becomes ever more trivial in content. And finally, there is the phenomenon of the rise of religious and political fanaticism and terrorism opposed, it can seem, either in a faint-hearted and self-doubting way, or brutally by war and the suspension of justice, apparently confirming Yeats's lines "The best lack all conviction, while the worst are full of passionate intensity".

Most serious of all, there is the impending crisis of global warming. There is the real possibility that average global temperature will rise by 3 to 6 or even 10 degrees centigrade by the end of the century, rendering vast tracts of the earth's surface, at present densely populated, uninhabitable, sea levels rising by a meter or so, flooding many great cities of the world. Reports from experts about the pace of global warming – shrinking of ice at the poles, contraction of glaciers – grow steadily more alarming year by year.

We have known about global warming for a long time. John Tyndall discovered that carbon dioxide is a greenhouse gas as long ago as 1859, and Svante Arrhenius realized in 1896 that we would cause global warming. Living in Sweden, he thought it would be a good thing. But the first person really to discover that we are *causing* global warming was Guy Callendar, who gave a lecture to the Meteorological Society in London on the subject in 1938. He was not believed. Of course, 1938 was not the best time to make the announcement! Any lingering doubts should have been removed, however when, in the early 1960s, Charles Keeling made extremely accurate measurements of the increase in carbon dioxide in the atmosphere.[2]

What is so shocking is that it has taken so long – several decades – for humanity to begin to take the impending threat seriously; let alone work out what needs to be done; let alone do it.

Global warming threatens to intensify all our other global problems – apart, perhaps, from that of rapid population growth (which might be curtailed by starvation, floods, drought, and war, all provoked by global warming).

If we are to realize what is genuinely of value to us in life more successfully than we have in the past we must, at the very least, discover how to resolve these immense global problems in very much more humane, intelligent, and effective ways than we have managed to do so far.

The Role of Modern Science and Technology
Modern science and technology have made immense contributions to the enrichment of human life. The modern world

---

[2] Weart (2003).

is inconceivable without them. But they have also made possible all our current global problems. Modern science and technology make possible modern medicine and hygiene, modern agriculture and industry which, in turn, have led to population growth, destruction of natural habitats and rapid extinction of species. Modern science, technology and industry being developed in some countries, but not in others, have led to immense differences in wealth around the world. Science and technology have made modern armaments possible, and the lethal character of modern warfare. As a result, the more scientifically advanced countries have been able to impose their will on those without modern science. Even aids is spread by modern methods of travel, made possible by modern technology. And of course global warming is a product of modern industry and agriculture, made possible by modern science and technology.

It is not just that modern science has made these things *possible*. In a perfectly respectable sense of "cause", all our global problems have been *caused* by modern science and technology.

It may be objected that it is not *science* that is the cause of these global problems but rather the things that we *do*, made possible by science and technology. This is obviously correct. But it is also correct to say that scientific and technological progress *is* the cause. The meaning of "cause" is ambiguous. By "the cause" of event E we may mean something like "the most obvious observable events preceding E that figure in the common sense explanation for the occurrence of E". In this sense, human actions (made possible by science) are the cause of such things as people being killed in war, destruction of tropical rain forests. On the other hand, by the "cause" of E we may mean "that prior change in the environment of E which led to the occurrence of E, and without which E would not have occurred". If we put the 20th century into the context of human history, then it is entirely correct to say that, in this sense, scientific-and-technological progress is the cause of our distinctive current global disasters: what has changed, what is new, is scientific knowledge and technological know-how, not human nature. Give a group of chimpanzees rifles and teach them how to use them and in one sense, of course, the cause of the subsequent demise of the group would be the actions of the

304

chimpanzees. But in another obvious sense, the cause would be the sudden availability and use of rifles – the new, lethal technology. Yet again, from the standpoint of theoretical physics, "the cause" of E might be interpreted to mean something like "the physical state of affairs prior to E, throughout a sufficiently large spatial region surrounding the place where E occurs". In this third sense, the sun continuing to shine is as much a part of the cause of war and pollution as human action or human science and technology.

In short, if by the cause of an event we mean that prior change which led to that event occurring (the second of the above three senses), then it is the advent of modern science and technology that has *caused* all our current global crises. It is not that people became greedier or more wicked in the 19th and 20th centuries; nor is it that the new economic system of capitalism is responsible, as some historians and economists would have us believe.[3] The crucial factor is the creation and immense success of modern science and technology. This has led to modern medicine and hygiene, to population growth, to modern agriculture and industry, to habitat loss and rapid extinction of species, to pollution of land, sea and air, to world wide travel (which spreads diseases such as aids), to global warming, and to the destructive might of the technology of modern war and terrorism, conventional, chemical, biological and nuclear.

It is tempting to blame modern science and technology for our troubles. But that misses the point. We need modern science and technology, to help us know what our problems are, and to help us solve them. We would not know we were causing global warming without modern science (even if there would be no global warming if there were no science). The fault lies, not with science *per se*, but rather with scientific and technological research *dissociated from the more fundamental quest to discover how to help humanity solve its global problems and make progress towards as good a*

---

[3] Science plus communism would have done the trick just as well – even better, in fact, as the record of the Soviet Union reveals (in connection with environmental degradation, for example).

world as possible. The fault lies with our long-standing failure to pursue science within the framework of wisdom-inquiry.[4]

As I argued in chapter 6, we face two great problems of learning: learning about the universe, and ourselves as a part of the universe; and learning how to create civilization.[5] We have solved the first problem. The solution is science. But we have failed to solve the second problem. And that, inevitably, puts us into a situation of unprecedented danger. For the solution to the first problem – science – bequeaths to us unprecedented power to act which, without civilization or wisdom, is almost bound to do as much harm as good. Just this has been the outcome.

The crisis of all crises, in short – the crisis behind all the others – is science without wisdom or, more accurately, science without wisdom-inquiry.

As a matter of supreme urgency, we need to bring about a revolution in academia so that it takes up its proper task of devoting reason – the genuine article – to the pursuit of wisdom, to helping us tackle and solve our immense, intractable global problems.

---

[4] Or, put in the terms of this book, the fault lies with our long-standing failure to take, as our fundamental problem, to help the God-of-Cosmic-Value to flourish in the God-of-Cosmic-Power. This is our fundamental problem of living, and also our fundamental *intellectual* problem. If this were understood, it would be obvious that wisdom-inquiry is what we require to help us improve our attempts at solving this fundamental problem. It is our failure to appreciate that this *is* our fundamental problem which has made it possible to dissociate science from religion, from concern with what is of value in existence, and in turn made it possible to develop social inquiry as social *science* (the pursuit of knowledge of social phenomena), and not as the endeavour to help humanity realize what is of value in life.

[5] Both are subordinate aspects of our fundamental problem of helping the God-of-Cosmic-Value to flourish within the God-of-Cosmic-Power. In the last analysis, our current global problems are the outcome of our failure to understand that this is our fundamental problem – our failure to give it priority in personal, social, economic and political life.

What Do We Need to Do?

What do we need to do to solve our global problems? I now indicate very briefly what in my view needs to be done, taking the main problems in turn.

Global Warming. This would seem to be the most serious of our problems. Let me state the obvious. In order to come to grips with this problem, the industrially advanced world needs to cut back on its emissions of CO_2 as rapidly as possible. We must stop burning oil and coal, and rapidly develop alternative sources of power: wind, hydro-electric, wave, tidal, sunlight via photoelectric cells, biomass fuels and, perhaps, nuclear power. Vehicles powered by petrol must be replaced by vehicles powered by batteries (charged by electricity in turn produced by sustainable means that do not emit CO_2). Energy saving devices need to be installed in homes, offices, factories and other buildings. Street lighting needs to be made more energy efficient. At the same time, global cooperation is required to put an end to the destruction of tropical rain forests, which significantly contributes to global warming.

Many of these measures are highly problematic, for both technical and social reasons. Wind power, hydro-electric power, and tidal power all tend to have adverse environmental consequences. Growing biomass fuels takes land away from the production of crops for much needed food. Nuclear power is, of course, notoriously problematic, in part because of the long-lasting, highly radioactive material that it produces, in part because of the link with nuclear weapons. Electric vehicles at present have nothing like the range or power of petrol or diesel fuelled vehicles. It is not clear what is to replace oil when it comes to ships, and aeroplanes

It may prove possible to harvest sunlight on an industrial scale by means of photo-electric panels spread over square miles in deserts. But photo-electric panels are expensive, and there are problems of transporting electricity to cities and densely populated areas – which tend to be far away from deserts.

There are speculative ideas about how it might be possible to extract CO_2 from the atmosphere in sufficient quantities to make a difference, or to cut down on the amount of sunlight reaching the earth, for example by sending mirrors into space between us and

307

the sun. All these ideas seem at present impractical, because of expense or adverse consequences or, quite simply, because they would not work.

The world needs to cooperate on putting a stop to the destruction of tropical rain forests. Countries such as Brazil and Indonesia need financial and other assistance from the industrially advanced world. Tropical rain forests require international policing to stop destructive logging.

The planet will continue to grow warmer even if we stopped all emissions of CO_2 overnight. This is because there is a delay in the planetary system. The CO_2 we have already put into the atmosphere will continue to turn up the heat for some time to come. As it is, of course, it will at best take decades for the world to reduce substantially its emissions of CO_2. Global warming will continue for decades to come. Low lying islands and coastal regions will have to be abandoned, as sea levels rise, and other regions will have to be abandoned because of heat and drought. As populations rise, land available for habitation and agriculture will shrink, not a good prospect for peace. World-wide cooperation will be needed to take care of refugees who come from regions made uninhabitable by global warming.

War. The world needs an international peace-keeping force which can be deployed swiftly anywhere on earth to intervene if violent conflict seems likely, or has already broken out, whether internal to a country, or between nations. At present, the UN is supposed to perform this function, but does so ineffectually, partly because it cannot intervene in civil war, partly because the UN security council must reach agreement, and this is either not forthcoming at all, or only after a protracted period of wheeling and dealing. Sometimes the UN supports military intervention it ought not to support, as in the case of the Afghanistan war[6] after 9/11, while on

---

[6] 9/11 was a monstrous crime, not an act of war, and could not conceivably justify war in retaliation. The UN issued a resolution which in effect supported the USA in its subsequent invasion of Afghanistan. It did so, in my view, because the aggrieved nation was the USA. If, instead, France had been the victim, the Louvre being destroyed in an analogous terrorist attack with, we may suppose, a similar loss of life (around 3,000 people), I feel sure the UN would not have supported

other occasions it fails to support intervention it clearly ought to support, as in cases of conflict in Africa, in the former Yugoslavia, and in Rwanda.

In order to have an international peace-keeping force that does the job properly, we probably first need to establish a democratic, enlightened world government. That, it might be argued, rather puts the cart before the horse. We will only be able to establish a democratic world government if we have already established world peace. It seems reasonable to hold, however, that efforts to establish world peace should work in tandem with efforts to establish democratic world government.

More than an effective, humanitarian peace-keeping force is required to establish world peace, as the case of Europe graphically illustrates. For centuries, Europe suffered war after war, culminating in the horrors of the first and second world wars, both of which had their source in Europe. After the second world war, a number politicians and others worked hard to develop trade and other interconnections between European states such that all future European wars would be unthinkable. This hope has been fully realized. Yugoslavia does not really constitute an exception since that country was never a part of the efforts to create the Common Market, or the European Union. We have here something like a model for what we should try to create world-wide. For this to succeed, though, it will probably be necessary for there to be democracies in all the counties of the world, and far greater equality of wealth than at present around the world. (This proposal is very definitely not the view that the rest of the world should become European in character and culture; it is rather the view that something important is to be learned from the manner in which European peace has been established after centuries of war, for the establishment of peace throughout the rest of the world. We have here a particular example of what can be accomplished.)

We require, too, a massive reduction in armaments and the military, all over the world, and especially in the USA and UK. All nuclear weapons need to be destroyed, and the arms industry needs to be massively curtailed.

France in a retaliatory invasion of Afghanistan.

Population Growth. The world's population is predicted to rise to over 9 billion by 2050. Population growth adds to global warming, increases likelihood of war, undermines economic growth, and tends to speed up destruction of natural habitats, extinction of species, and over fishing of the sea. One relatively cheap and practical measure that could be taken to slow down population growth would be to ensure that every woman on the planet of child bearing age has access to reliable birth control methods: the pill, the condom, the coil. It does not help that this is opposed by the Catholic Church, and was opposed by the Bush administration in the USA. One view is that population growth tends to level off as countries become wealthier. Parents tend not to have so many children – the argument goes – because the need to provide them with an education makes children more expensive, parents do not need to have children to care for them in old age because they can rely on state care, and falling death rates among children mean that it is no longer seen as essential to have lots of children to ensure that some survive. It is foolish to rely on these mechanisms, however, to slow down population growth. What is required is an effective programme world wide to ensure that every woman of child bearing age has access to reliable contraception.

World Poverty. The debt of the poor countries of the world needs to be cancelled. There needs to be a change in world trading agreements, to ensure that it is the poor countries that are favoured, and not the rich. It must be permitted for poor countries to implement protectionism, to protect fledging industries against international competition.

A new global Marshall Plan needs to be created, funded by the wealthy countries of the world – the USA, Canada, Europe, Japan, Australia, New Zealand, and perhaps others – to help poor countries develop in as sustainable a way as possible, the emphasis being on education and the development of appropriate industry and agriculture. This needs to be allied to efforts to promote democracy, and to put a stop to political corruption. More scientific and technological research needs to be devoted to the problems of the poor: problems of health, agriculture, communications, education, appropriate industrial development.

Destruction of Natural Habitats and Extinction of Species. As an integral part of the global Marshall plan, indicated above, wealthy countries need to collaborate with poor and developing countries to take those measures required to stop the destruction of tropical rain forests and other natural habitats. This involves both deploying and adequately financing and equipping environmental police to put a stop to logging and hunting. It also involves providing aid for alternative, more sustainable methods of development. Agriculture needs to be developed in such a way that habitats remain for wild life to flourish. There needs to be enhanced protection for endangered species.

I put these global policy proposals forward, not because I think they make a startlingly original contribution to thought about how we are to solve our global problems, but rather to indicate the kind of things we need to do to solve these problems. We need this as background to help answer the crucial question of the next section: "How would wisdom-inquiry help us put global policies such as these successfully into practice?"

I am well aware that some governments, many NGOs, the UN, social businesses, countless individual and officials are already working hard to implement many aspects of these policies. Despite all these efforts, progress towards implementing the policies I have indicated (or better versions of these policies) remains agonizingly slow. Some of our global problems are intensifying – most notably global warming.

Some may complain that not enough detail has been given to assess these policy proposals. I have, however, I think, said enough for the purposes of the argument of the next section. Others may complain that some, or even all, of what I have proposed is wrong-headed, and such that, if put into practice, would have dire consequences, the very opposite of what is intended. Those who believe in the universal efficacy of the free market to solve our problems are likely, in particular, to object to much of the above. My reply is that even if the above policies are misguided, in part or in total, this will not substantially affect the argument of the next section. It must be remembered that a basic task of wisdom-inquiry is (a) to articulate global problems, and (b) propose and critically assess possible solutions. Nothing is presupposed about what our problems are, and what we need to do about them: wisdom-inquiry is intended to help enlighten us about these matters. Furthermore, even if we do need different policies from the above to solve our problems, nevertheless the argument of the next section goes through – as long as it is agreed that we need to tackle our problems *democratically*.

How Could Wisdom-Inquiry Help?

How exactly, it may be asked, could wisdom-inquiry help humanity implement these policies – if that is what is required – and thus help solve our global problems in a way which is so much more effective than knowledge-inquiry? Let us suppose that the academic revolution has occurred. Universities everywhere put wisdom-inquiry into practice. How could this make such a substantial difference to our capacity to solve global problems humanely and effectively, thus making progress towards as good a world as possible?

In essence, the answer is extremely simple. Our only hope of solving our global problems successfully lies with tackling these problems *democratically*. Benevolent, enlightened dictatorships or autocracies will not meet with success. But if democratic tackling of global problems is to succeed, we first need democracy to be established around the world, and second we need electorates – the world's population – to have an enlightened understanding of what our global problems are, and what we need to do about them. If this is lacking, democratic governments will not be able to implement the policies that are required. If, on the other hand, a majority of the world's people do have a good understanding of what our problems are, and what needs to be done about them, there is a good chance governments will respond to what this majority demands. This assumes, of course, that it is in the interests of the majority that global problems be solved. If this is not the case, then many might see clearly what needs to be done, but might nevertheless oppose the doing of it. I shall discuss this possibility in the next but one section.

A crucial requirement for tackling global problems successfully, then, is that a majority of the world's people have a good understanding of what these problems are, and what needs to be done about them. This is quite drastically lacking at present. Indeed, it may seem quite absurdly utopian to think it would ever be possible for most people on earth to agree about what our problems are, and what we need to do about them.

Step forward wisdom-inquiry. It is just here that wisdom-inquiry makes a dramatic difference. A basic task of wisdom-inquiry is (a) to articulate problems of living, including global problems, and (b) to propose and critically assess possible solutions – actual and possible actions, policies, political programmes, economic strategies, philosophies of life.[7]

A university that puts wisdom-inquiry into practice would hold a big Seminar once a month (let us say) devoted to discussing what our global problems are, and how they are to be solved. Everyone at the university would be invited to attend and participate, from undergraduate to professor and vice-chancellor. The Seminar might sometimes be big affairs, involving the media, with well-known speakers, while on other occasions it might be smaller, more private, an affair for a group of specialists, devoted to some specific issue. The aim would be, not just to highlight existing problems, or criticize existing policies, but to come up with workable, realistic, effective new policies. The constitution of the university would be such that good ideas developed in the Seminar would be capable of influencing more specialized research in the university, and would be critically assessed by such research. One result of the Seminar would be that all those associated with, and educated in, universities, from professor to undergraduate, would acquire a good understanding of what our global problems are, what is and is not being done about them, what could be done, and what kind of research and education is required to help solve them. A long-term task of social inquiry would be to help build aim-oriented rationality into our diverse institutions – government, industry, finance, agriculture, international trade, the military, the

---

[7] Even if the policies I have outlined are the best available, they need to be developed in far greater detail before they qualify even for serious consideration. The chances are, of course, that what I have proposed deserves to be rejected, because it is unworkable, undesirable, or both.

media, the law, education – so that problematic aims may be transformed to become those that help solve global problems. A fundamental task for universities implementing wisdom-inquiry is to educate the public about what our global problems are, and what we need to do about them. This would be done, not by *instruction*, but by lively discussion and debate, ideas, arguments and information flowing in both directions. There would be powerful inducements for academics to engage in public education by means of public discussion and lectures, articles in newspapers, popular books, broadcasts, blogs on the internet, even novels and plays. All academics want to make a contribution to academic thought, not only for its own sake, but also because this leads to academic status and prestige, academic prizes, and career advancement. Granted wisdom-inquiry, contributions are judged in terms of their capacity to help people realize what is of value in life. Working within the framework of wisdom-inquiry, academics would, in other words, be highly motivated to engage in the kind of public education I have indicated (since this is integral to what counts as an academic contribution). A central purpose of academia would be to promote cooperatively rational tackling of problems of living in the social world, and put aim-oriented rationality into practice in personal and social life. The problematic aims and priorities of scientific and technological research would be subjected to sustained, imaginative exploration and criticism, by academics and non-academics alike, this feeding into, and making use of, the discussion of problems of living going on within and without academia. Wisdom-inquiry is designed to engage in rational discussion of political policies and programmes, and to promote this as well. Universities would have just sufficient power to retain their independence from pressures of government, public opinion, industry, and the media, but no more. It would be standard for a nation's universities to include a shadow government. If the actual government does not permit such a thing, universities would clamour to be free to create it and, in doing so, and would receive international support. The nation's university shadow government would be entirely without power, but would also be free of all the constraints and pressures that actual power is subject to, which tend to distort and corrupt what actual governments do. The shadow university government would seek to develop and publish ideal possible actions, policies and legislative programmes which the nation's actual government ought to be developing and enacting. The idea would be that learning would go on in both directions, the ideal university shadow government learning about the realities of power, the nation's actual government learning to distinguish what is merely politically expedient from what is in the interests of the nation and humanity, a fund of good ideas for policies and legislation being readily available from the shadow government. Finally, the world's universities would contain a shadow university world government which would do, for the world, what national shadow governments do for nations. A basic task would be to work out how an actual world government might be created, what form this should take, what its desirable and undesirable consequences would be likely to be.

In brief, the whole character, structure, activity, aims and ideals of wisdom-inquiry universities would be such as to be devoted to helping humanity learn how to resolve global problems in increasingly cooperatively rational ways, thus making increasingly assured progress towards as good a world as possible. Universities would be humanity's means to learn how to create a genuinely civilized world

The contrast with knowledge-inquiry is devastating. Knowledge-inquiry fails to do almost everything that needs to be done to help humanity make progress in tackling global problems. Knowledge-inquiry does, it is true, acquire knowledge and technological know-how, and make this available, primarily to government agencies and industry, to be used to solve practical problems. This can undeniably be of great value and, as we have seen, has made possible the creation of the modern world. But almost everything else that needs to be done is rigorously excluded from the intellectual domain of academia under the misguided idea that this is necessary to preserve the objectivity and reliability, the authentically factual character, of the knowledge that is acquired. Far from giving priority to (a) articulating global problems, and (b) proposing and critically assessing possible solutions, these vital intellectual activities are excluded from knowledge-inquiry altogether, on the grounds that they involve politics, values, action,

315

human suffering, morality, and can only undermine, and not contribute to, the pursuit of factual knowledge. Again, far from giving priority to the task of introducing aim-oriented rationality into the social world, knowledge-inquiry does not even put aim-oriented rationality into practice *itself*, in science, social inquiry or the humanities. There is no place for the Seminar devoted to tackling global problems. Social science and the humanities seek to improve knowledge and understanding of social and cultural phenomena, but do not actively try to transform social life. Individual academics may take it upon themselves to contribute to public education but this is, as it were, an extra-curriculum activity, not a part of the official business of professional academic life – which is to contribute to the growth of knowledge. Far from academia encouraging discussion and debate with the public, ideas being encouraged to flow in both directions, knowledge-inquiry, quite to the contrary, demands that the intellectual domain of inquiry be sealed off from the corrupting influence of the social world, so that only those considerations relevant to the acquisition of knowledge of truth may influence what is accepted and rejected – such as evidence and valid argument. Knowledge-inquiry provides every inducement to academics to seek to contribute to knowledge, but no inducement whatsoever to engage in the extra-curriculum activity of public education (since this does not contribute to knowledge). What matters is how well-established and significant a contribution to knowledge is, not whether it does, or does not, help enhance the quality of human life. The intellectual standards of knowledge-inquiry are almost exclusively concerned with the problem of distinguishing authentic contributions to knowledge from would-be contributions that fail to pass master, in one way or another. These standards are not concerned to help improve the aims and priorities of research. Choosing what research aims receive financial support, and what do not, is left to research funding bodies to decide: it is not thrown open to sustained scientific and public discussion and debate. Inevitably, as a result, research priorities come to reflect the interests of those who do science, and those who pay for it – government and industry – rather than the interests of those whose needs are the greatest, the poor of the earth who, being poor, do not

316

have the means to pay for scientific research. Vast sums are spent on military research, very little in comparison on research related to the diseases and problems of the poor of Africa, south America and Asia. Finally, there can be no place for a shadow government in the university, granted knowledge-inquiry. Politics is to be excluded altogether from the intellectual domain of inquiry; only the pursuit of knowledge about political life is permitted.

The outcome of this wholesale failure to do what most needs to be done, apart from acquire knowledge, is just what might be expected. Much knowledge is acquired but this, in the absence of a more fundamental concern to help humanity solve global problems, does as much harm as good. Knowledge-inquiry, *instead of helping to solve global problems, helps to create and intensify them*, as we have seen.

I have concentrated on *universities*. But if the revolution were to occur in universities, it would have an impact throughout the whole educational and research world, as well as influencing dramatically, as I have tried to indicate, the media, government, the arts, the law, industry, agriculture, international relations, and personal and social life quite generally.

Changing knowledge-inquiry into wisdom-inquiry in universities throughout the civilized world would make a dramatic difference to the capacity of humanity to tackle global problems successfully.

Objections

Objection 1: Academics would never agree to put wisdom-inquiry into practice.

Reply: The arguments for the greater rationality, intellectual integrity and potential human value of wisdom-inquiry are overwhelming. Once these arguments have been understood by a sufficient number of influential academics, funding bodies and university administrators, universities will begin to move piecemeal towards wisdom-inquiry. Indeed, as I shall show in the next section, this academic transformation is, to some extent, already underway.

Objection 2: Governments, industry, public opinion would never permit the required academic revolution to take place.

317

Reply: Undoubtedly in some parts of the world today it would indeed be impossible. There would be difficulties in North Korea, Burma, Zimbabwe, Saudi Arabia, Iran, and even China and Russia. Even in the 30 full democracies of the world,[8] serious attempts to instigate wisdom-inquiry would meet with opposition. Even democratically elected governments are unlikely to take kindly to academic criticism of their policies, and to the creation of academic shadow governments. Those universities that took a lead in implementing wisdom-inquiry might find they were being penalized by having government funding decreased. Industry might withdraw funds as well. Academia would have an incredibly powerful argument in its hands to combat such manoeuvres: the changes are needed in the interests of rationality, intellectual integrity, and the future of humanity. The public could be alerted to the scandal of government attempting to suppress academic thought devoted to helping humanity make progress towards as good a world as possible. This objection does not look very plausible when one takes into account that the academic revolution, from knowledge to wisdom, is already underway to some extent, in the UK and elsewhere, as we shall see in the next section.

Objection 3: Even if the academic revolution occurred, it would have little impact, either because academics failed to agree among themselves, or because they are ignored by centres of power and influence.

Reply: A nightmare possibility is that wisdom-inquiry academics simply reproduce all the standard ideas, prejudices and disagreements of the social world around them. In the US, academics supporting the Democrats might slug it out with those supporting the Republicans, and no one learns anything. I acknowledge that this is a possibility, but it would betray the fundamental intellectual ideals of wisdom-inquiry. Those engaged in social inquiry need to treat policy ideas in a way that is

---

[8] *The Economist* has recently assessed the democratic character of the countries of the world: see http://en.wikipedia.org/wiki/Democracy_Index There are 51 dictatorships, with North Korea at the bottom of the list.

analogous, in important respects, to the way natural scientists treat scientific theories: some such ideas may be hopeless, others may be partly good, partly bad, none is likely to be entirely good and sound, the all-important point is to pick out the best idea from its rivals, and subject it and its rivals to sustained critical examination, taking experience into account where possible, and if a better idea emerges from the pool of rivals, that should be adopted instead. It is of course just this that aim-oriented rationality is designed to facilitate, in the field of ideas for solutions to problems of living, on analogy with what aim-oriented empiricism facilitates within natural science. It will, for many reasons, be more difficult to protect wisdom-inquiry social thought from subversion than it is to protect natural science from subversion. Policy ideas implicate our lives, passions, ideals and values directly, and are much harder to assess rationally and by means of experience, than are scientific ideas. Experiments in the social world cannot be conducted freely in the way in which scientific experiments can.

As for academia being ignored even if it comes up with excellent, agreed ideas this, to some extent, is almost bound to occur. But only to some extent, and for a time. It took scientists decades to get governments, industry, the media and the public to take global warming seriously. The long-standing failure to get the message across has finally led scientists to make changes to the nature of science – nudging things towards wisdom-inquiry, as we shall see in the next section. But finally, at the time of writing (2009), the message has been delivered although there are few signs, as yet, that much is being done to reduce CO_2 emissions, in response to this message. In my view, the global warming message would have been communicated two or three decades earlier if wisdom-inquiry had been in place by 1945, let us say. The academic revolution we are considering would undoubtedly have a major impact, in the ways I have indicated, even if this impact would not be felt overnight, but would take a decade or so to filter through the intricacies of the social world.

Objection 4: Even if the academic revolution occurred, even if it came up with excellent policies and technologies, and even if these were appreciated and understood by governments and public alike, still this would not make much difference because the barrier to solving global problems is not lack of knowledge and understand, but the unwillingness of the wealthy to make the necessary sacrifices. Too many wealthy, powerful people do not want to do what needs to be done.

Reply: The policies I have indicated above would undoubtedly meet with resistance, were they ever to be seriously on the political agenda. In the USA, for example, business corporations are very good at protecting what they see as their interests by lobbying, by funding sympathetic politicians and political parties, and by manipulating the media. Even here, however, wisdom-inquiry could be effective, in that the public needs to become more enlightened about what these strategies are, and what needs to be done to combat them. This assumes that it is primarily the business and financial world which would want to oppose the policies we require. It could be argued that a majority of people living in wealthy countries do not want to support measures required to deal with global warming, or world poverty, because of the sacrifices that would have to be made. This, I believe, overestimates the sacrifices that are required, and underestimates concern people have for the future of the world. If policies are widely understood to be necessary, and likely to be effective, in tackling global warming, for example, or world poverty, and then I believe a majority of people in wealthy countries would be willing to endorse these policies, even if some sacrifice is required. Why should a global Marshall plan today meet with so much more resistance than the original Marshall plan encountered when first instigated after the second world war, when the USA was not as wealthy as it is today?

Is the Academic Revolution Underway?

So far I have drawn a stark contrast between knowledge-inquiry and wisdom-inquiry, and have suggested that knowledge-inquiry is at present dominant in universities all over the world. But is this really the case?

I have no doubt that it was the case 25 years ago. In 1983, for the first edition of my book *From Knowledge to Wisdom* I investigated six relevant aspects of academia to see which conception of inquiry prevailed, and found that knowledge-inquiry

320

was overwhelmingly dominant.[9] However, more recently, in 2003, I repeated the survey for the second edition of the book, and found that some changes had taken place in the direction of wisdom-inquiry, although knowledge-inquiry still dominated.[10] Since 2003, there have been further developments that have nudged some universities in the direction of wisdom-inquiry.

It is possible that the academic revolution really is underway, and we are in the middle of a dramatic transition from knowledge-inquiry to wisdom-inquiry. I now indicate some developments that have taken place in universities in the UK during the last twenty years which can, perhaps, be interpreted as constituting steps towards wisdom-inquiry.[11]

Perhaps the most significant steps towards wisdom-inquiry that have taken place during the last twenty years are the creation of departments, institutions and research centres concerned with social policy, with problems of environmental degradation, climate change, poverty, injustice and war, and with such matters as medical ethics and community health. For example, a number of departments and research centres concerned in one way or another with policy issues have been created at my own university of University College London during the last 20 years.

At Cambridge University, there is a more interesting development. One can see the first hints of the institutional structure of wisdom-inquiry being superimposed upon the existing structure of knowledge-inquiry. As I have indicated, wisdom-inquiry puts the intellectual tackling of problems of living at the heart of academic inquiry, this activity being conducted in such a way that it both influences, and is influenced by, more specialized research. Knowledge-inquiry, by contrast, organizes intellectual activity into the conventional departments of knowledge: physics, chemistry, biology, history and the rest, in turn subdivided, again and again, into ever more narrow, specialized research disciplines. But this knowledge-inquiry structure of ever more specialized research is hopelessly inappropriate when it comes to tackling our

---

[9] See my (1984), ch. 6.
[10] See my (2007a), ch. 6.
[11] What follows is adapted from my (2009).

major problems of living. In order to tackle environmental problems, for example, in a rational and effective way, specialized research into a multitude of different fields, from geology, engineering and economics to climate science, biology, architecture and metallurgy, needs to be connected to, and coordinated with, the different aspects of environmental problems. The sheer urgency of environmental problems has, it seems, forced Cambridge University to create the beginnings of wisdom-inquiry organization to deal with the issue. The "Cambridge Environmental Initiative" (CEI), launched in December 2004, distinguishes seven fields associated with environmental problems: conservation, climate change, energy, society, water waste built environment and industry, natural hazards, society, and technology, and under these headings, coordinates some 102 research groups working on specialized aspects of environmental issues in some 25 different (knowledge-inquiry) departments: see http://www.cei.group.cam.ac.uk/ . The CEI holds seminars, workshops and public lectures to put specialized research workers in diverse fields in touch with one another, and to inform the public. There is also a CEI newsletter.

A similar coordinating, interdisciplinary initiative exists at Oxford University. This is the School of Geography and the Environment, founded in 2005 under another name. This is made up of five research "clusters", two previously established research centres, the Environmental Change Institute (founded in 1991) and the Transport Studies Institute, and three inter-departmental research programmes, the African Environments Programme the Oxford Centre for Water Research, and the Oxford branch of the Tyndall Centre (see below). The School has links with other such research centres, for example the UK Climate Impact Programme and the UK Energy Research Centre.

At Oxford University there is also the James Martin 21st Century School, founded in 2005 to "formulate new concepts, policies and technologies that will make the future a better place to be". It is made up of fifteen Institutes devoted to research that ranges from ageing, armed conflict, cancer therapy and carbon reduction to nanoscience, oceans, science innovation and society, the future of the mind, and the future of humanity. At Oxford there is also the

Smith School of Enterprise and the Environment, founded in 2008 to help government and industry tackle the challenges of the 21st century, especially those associated with climate change.

Somewhat similar developments have taken place recently at my own university, University College London. Not only are there 141 research institutes and centres at UCL, some only recently founded, many interdisciplinary in character, devoted to such themes as ageing, cancer, cities, culture, public policy, the environment, global health, governance, migration, neuroscience, and security. In addition, very recently, the attempt has been made to organize research at UCL around a few broad themes that include: global health, sustainable cities, intercultural interactions, and human wellbeing. This is being done so that UCL may all the better contribute to solving the immense global problems that confront humanity.

All these developments, surely echoed in many universities all over the world, can be regarded as first steps towards implementing wisdom-inquiry.

Equally impressive is the John Tyndall Centre for Climate Change Research, founded by 28 scientists from 10 different universities or institutions in 2000. It is based in six British universities, has links with six others, and is funded by three research councils, NERC, EPSRC and ESRC (environment, engineering and social economic research). It "brings together scientists, economists, engineers and social scientists, who together are working to develop sustainable responses to climate change through trans-disciplinary research and dialogue on both a national and international level – not just within the research community, but also with business leaders, policy advisors, the media and the public in general" (www.tyndall.ac.uk/general/about.shtml). All this is strikingly in accordance with basic features of wisdom-inquiry.[12] We have here, perhaps, the real beginnings of wisdom-inquiry being put into academic practice.

A similar organization, modelled on the Tyndall Centre, is the UK Energy Research Centre (UKERC), launched in 2004, and also funded by the three research councils, NERC, EPSRC and ESRC.

---

[12] Tyndall Centre (2006).

Its mission is to be a "centre of research, and source of authoritative information and leadership, on sustainable energy systems" (www.ukerc.ac.uk/). It coordinates research in some twelve British universities or research institutions. UKERC has created the National Energy Research Network (NERN), which seeks to link up the entire energy community, including people from academia, government, NGOs and business.

Another possible indication of a modest step towards wisdom-inquiry is the growth of peace studies and conflict resolution research. In Britain, the Peace Studies Department at Bradford University has "quadrupled in size" since 1984 (Professor Paul Rogers, personal communication), and is now the largest university department in this field in the world. INCORE, an International Conflict Research project, was established in 1993 at the University of Ulster, in Northern Ireland, in conjunction with the United Nations University. It develops conflict resolution strategies, and aims to influence policymakers and others involved in conflict resolution. Like the newly created environmental institutions just considered, it is highly interdisciplinary in character, in that it coordinates work done in history, policy studies, politics, international affairs, sociology, geography, architecture, communications, and social work as well as in peace and conflict studies. The Oxford Research Group, established in 1982, is an independent think tank which "seeks to develop effective methods whereby people can bring about positive change on issues of national and international security by non-violent means" (www.oxfordresearchgroup.org.uk/). It has links with a number of universities in Britain. Peace studies have also grown during the period we are considering at Sussex University, Kings College London, Leeds University, Coventry University and London Metropolitan University. Centres in the field in Britain created since 1984 include: the Centre for Peace and Reconciliation Studies at Warwick University founded in 1999, the Desmond Tutu Centre for War and Peace, established in 2004 at Liverpool Hope University; the Praxis Centre at Leeds Metropolitan University, launched in 2004; the Crime and Conflict

Centre at Middlesex University; and the International Boundaries Research Unit, founded in 1989 at Durham University.[13]

Additional indications of a general movement towards aspects of wisdom-inquiry are the following. Demos, a British independent think tank has, in recent years, convened conferences on the need for more public participation in discussion about aims and priorities of scientific research, and greater openness of science to the public.[14] This has been taken up by The Royal Society which, in 2004, published a report on potential benefits and hazards of nanotechnology produced by a group consisting of both scientists and non-scientists. The Royal Society has also created a "Science in Society Programme" in 2000, with the aims of promoting "dialogue with society", of involving "society positively in influencing and sharing responsibility for policy on scientific matters", and of embracing "a culture of openness in decision-making" which takes into account "the values and attitudes of the public". A similar initiative is the "science in society" research programme funded by the Economic and Social Research Council which has, in the Autumn of 2007, come up with six booklets reporting on various aspects of the relationship between science and society. Many scientists now appreciate that non-scientists ought to contribute to discussion concerning science policy. There is a growing awareness among scientists and others of the role that values play in science policy, and the importance of subjecting medical and other scientific research to ethical assessment. That universities are becoming increasingly concerned about these issues is indicated by the creation, in recent years, of many departments of "science, technology and society", in the UK, the USA and elsewhere, the intention being that these departments will concern themselves with interactions between science and society.

Even though academia is not organized in such a way as to give intellectual priority to helping humanity tackle its current global problems, academics do nevertheless publish books that tackle these issues, for experts and non-experts alike. For example, in

---

[13] For an account of the birth and growth of peace studies in universities see Rogers (2006).

[14] See Wilsdon and Willis (2004).

recent years many books have been published on global warming and what to do about it: see: www.kings.cam.ac.uk/assets/d/da/Global _Warming_bibliography.pdf

Here are a few further scattered hints that the revolution, from knowledge to wisdom, may be underway – as yet unrecognized and unorganized. In recent years, research in psychology into the nature of wisdom has flourished, in the USA, Canada, Germany and elsewhere.[15] Emerging out of this, and associated in part with Robert Sternberg, there is, in the USA, a "teaching for wisdom" initiative, the idea being that, whatever else is taught – science, history or mathematics – the teaching should be conducted in such a way that wisdom is also acquired.[16] There is the Arete Initiative at Chicago University which has "launched a $2 million research programme on the nature and benefits of wisdom": see http://wisdomresearch.org/. There are two initiatives that I have been involved with personally. The first is a new international group of over 200 scholars and educationalists called Friends of Wisdom, "an association of people sympathetic to the idea that academic inquiry should help humanity acquire more wisdom by rational means": see www.knowledgetowisdom.org. The second is a special issue of the journal *London Review of Education*; of which I was guest editor, devoted to the theme "wisdom in the university". This duly appeared in June 2007 (vol. 5, no.2). It contains seven articles on various aspects of the basic theme. Rather strikingly, another academic journal brought out a special issue on a similar theme in the same month. The April-June 2007 issue of *Social Epistemology* is devoted to the theme "wisdom in management" (vol. 21, no. 2). On the 5th December 2007, History and Policy was launched, a new initiative that seeks to bring together historians, politicians and the media, and "works for better public policy through an understanding of history": see www.historyandpolicy.org/.

Out of curiosity, on 18 May 2009, I consulted Google to see whether it gives any indications of the revolution that may be underway. Here are the number of web pages that came up for

---

[15] See, for example, Sternberg (1990).
[16] See Sternberg et al., (2007).

various relevant topics: "Environmental Studies" 9,910,000; "Development Studies" 7,210,000; "Peace Studies" 529,000; "Policy Studies" 2,160,000; "Science, Technology and Society" 297,000; "Wisdom Studies" 5,510; "From Knowledge to Wisdom" 18,100; "Wisdom-Inquiry" 625. These figures do not, perhaps, in themselves tell us very much. There is probably a great deal of repetition – and Google gives us no idea of the intellectual quality of the departments or studies that are being referred to. One of the items that comes up in Google is Copthorne Macdonald's "Wisdom Page" – a compilation of "various on-line texts concerning wisdom, references to books about wisdom, information about organizations that promote wisdom", and including a bibliography of more than 800 works on wisdom prepared by Richard Trowbridge.

None of these developments quite amounts to advocating or implementing wisdom-inquiry (apart from the two I am associated with). One has to remember that "wisdom studies" is not the same thing as "wisdom-inquiry". The new environmental research organizations, and the new emphasis on policy studies of various kinds, do not in themselves add up to wisdom-inquiry. In order to put wisdom-inquiry fully into academic practice, it would be essential for social inquiry and the humanities to give far greater emphasis to the task of helping humanity learn how to tackle its immense global problems in more cooperatively rational ways than at present. The imaginative and critical exploration of problems of living would need to proceed at the heart of academia, in such a way that it influences science policy, and is in turn influenced by the results of scientific and technological research. Academia would need to give much more emphasis to the task of public education by means of discussion and debate. As I have stressed, our only hope of tackling global problems of climate change, poverty, war and terrorism humanely and effectively is to tackle them democratically. But democratic governments are not likely to be all that much more enlightened than their electorates. This in turn means that electorates of democracies must have a good understanding of what our global problems are, and what needs to be done about them. Without that there is little hope of humanity making progress towards a better world. A vital task for universities is to help educate the public about what we need to do to avoid – at the least – the worst of future possible disasters. Wisdom-inquiry would undertake such a task of public education to an extent that is far beyond anything attempted or imagined by academics today. There is still a long way to go before we have what we so urgently need, a kind of academic inquiry rationally devoted to helping humanity learn how to create a better world

Nevertheless, the developments I have indicated can be regarded as signs that there is a growing awareness of the need for our universities to change so as to help individuals learn how to realize what is genuinely of value in life – and help humanity learn how to tackle its immense global problems in wiser, more cooperatively rational ways than we seem to be doing at present. My own calls for this intellectual and institutional revolution may have been in vain. But what I have been calling for, all these years, is perhaps, at last, beginning to happen. If so, it is happening with agonizing slowness, in a dreadfully muddled and piecemeal way. It urgently needs academics and non-academics to wake up to what is going on – or what needs to go on – to help give direction, coherence and a rationale to this nascent revolution from knowledge to wisdom.

Conclusion

The basic point is extremely simple. If we are to make better progress towards as good a world as possible, we need to learn how to do it. That in turn requires that we have in our hands institutions of learning rationally devoted to that task. It is just this that we do not have at present – although there are hints that such institutions might be struggling to be born. What we have at present is academic inquiry devoted to the pursuit of knowledge which, as we have seen, helps create as many problems as it solves. We urgently need to transform our universities so that they come to put wisdom-inquiry into practice. Only then will the God-of-Cosmic-Value, as it is represented on earth, flourish, embedded as it is within the God-of-Cosmic-Power.

BIBLIOGRAPHY

Abott, P. and C. Wallace, 1990, *An introduction to sociology: feminist perspectives*, Routledge, London.

Appleyard, B. 1992, *Understanding the Present: Science and the Soul of Modern Man*, Picador, London.

Ayer, A.J., 1936, *Language, Truth and Logic*, Gollancz, London.

Baldwin, J. M., 1896a, 'Heredity and Instinct: Discussion (Revised) following Professor C. Lloyd Morgan before the New York Academy of Sciences', *Science*, N.S. 3, pp. 438-41 and 558-61.

_____, 1896b, 'A New Factor in Evolution', *The American Naturalist*, 30, pp. 354-451 and 536-53.

Barnes, B. and D. Bloor, 1981, 'Relativism, Rationalism and the Sociology of Knowledge', in M. Hollis and S. Lukes, eds., *Rationality and Relativism*, Blackwell, Oxford, pp. 21-47.

Bartley, B., 1962, *The Retreat to Commitment*, Knopf, New York.

Bateson, P. 2004, 'The Active Role of Behaviour in Evolution', *Biology and Philosophy* 19, pp. 283-98.

Bell, J. S., 1987, *Speakable and unspeakable in quantum mechanics*, Cambridge University Press, Cambridge.

Berkeley, G., 1957, *A New Theory of Vision and Other Writings*, Everyman, London.

Berlin, I., 1979, *Against the Current*, Hogarth Press, London.

_____, 1999, *The Roots of Romanticism*, Chatto and Windus, London.

Berman, M., 1981, *The Reenchantment of the World*, Cornell University Press, Ithaca.

Bernanos, G., 1948, *Joy*, Bodley Head, London.

Bloor, D., 1976, *Knowledge and Social Imagery*, Routledge and Kegan Paul, London.

Bond, E.J., 1983, *Reason and Value*, Cambridge University Press, Cambridge.

Bonner, J., 1980, *The Evolution of Culture in Animals*, Princeton University Press, Princeton.

Bowler, P. J., 2009, *Evolution: The History of an Idea*, University of California Press, Berkeley.

Boyd, R. and P. Richerson, 1980, *Culture and the Evolutionary Process*, University of Chicago Press, Chicago.

_____, 2005, *Not By Genes Alone*, University of Chicago Press, Chicago.

Brink, D., 1989, *Moral Realism and the Foundations of Ethics*, Cambridge University Press, Cambridge.

Campbell, D. T., 1975, 'On the Conflicts Between Biological and Social Evolution and Between Psychology and Moral Tradition', *American Psychologist* 30, pp. 1103-26.

Churchland, P.M., 1994, 'Folk Psychology (2)' in *A Companion to the Philosophy of Mind*, ed. S. Gutenplan, Blackwell, Oxford, pp. 308-316.

Cronin, H., 1991, *The Ant and the Peacock*, Cambridge University Press, Cambridge.

D'Alembert, J., 1963, *Preliminary Discourse to the Encyclopedia of Diderot*, Bobbs-Merrill, New York (originally published in 1751).

Dampier, D., 1971, *A History of Science*, Cambridge University Press, Cambridge.

Dawkins, R., 1978, *The Selfish Gene*, Paladin, London.

_____, 2006, *The God Delusion*, Transworld, London.

Dennett, D., 1984, *Elbow Room*, Clarendon Press, Oxford.

_____, 1996, *Darwin's Dangerous Idea*, Penguin Books, London.

Descartes, R., 1949, *A Discourse on Method*, Everyman, London.

Dobzhansky, T., 1970, *Genetics of the Evolutionary Process*, Columbia University Press, New York.

Dobzhansky, T. and E. Boesiger, edited and completed by B. Wallace, 1983, *Human Culture: A Moment in Evolution*, Columbia University Press, New York.

Dunbar, R., 1996, *Grooming, Gossip and the Evolution of Language*, Faber and Faber, London.

Dunbar, R. *et al.*, 1999, *The Evolution of Culture*, Edinburgh University Press, Edinburgh.

Durant, J., 1997, 'Beyond the Scope of Science', *Science and Public Affairs*, Spring, pp. 56-57.

Durham, W., 1991, *Coevolution: Genes, Culture, and Human Diversity*, Stanford University Press, Stanford, California.

Einstein, A., 1949, 'Autobiographical Notes', in *Albert Einstein: Philosopher-Scientist*, ed., P. A. Schilpp, Open Court, La Salle, Illinois, pp. 3-94.

_____, 1973, *Ideas and Opinions*, Souvenir Press, London.

_____, 1986, *Letters on Wave Mechanics*, Philosophical Library, New York.

Fargaus, J., ed., 1993, *Readings in Social Theory*, McGraw-Hill, New York.

P. Feyerabend, P., 1978, *Against Method*, Verso, London.

_____, 1987, *Farewell to Reason*, Verso, London.

Fisher, R. A., 1930, *The Genetical Theory of Natural Selection*, Clarendon Press, Oxford.

Galileo, 1957, *The Assayer*, in *Discoveries and Opinions of Galileo*, ed. S. Drake, Doubleday, New York, pp. 229-280 (first published in Italian in 1623).

Gascardi, A., 1999, *Consequences of Enlightenment*, Cambridge University Press, Cambridge.

Gatti, H., 1999, *Giordano Bruno and Renaissance Science*, Cornell University Press, Ithaca, N.Y

Gay, P., 1973, *The Enlightenment: An Interpretation*, Wildwood House, London.

Ginzburg, C., 1980, *The Cheese and The Worms*, Routledge, London.

Goodall, J., 1971, *In the Shadow of Man*, Collins, London.

Grice, H.P., 1957, 'Meaning', *Philosophical Review*, 66, pp. 377-88.

_____, 1989, *Studies in the Way of Words*, Harvard University Press, Cambridge, Mass.

Gross, P., et al., 1996, eds., *The Flight from Science and Reason*, Annals of the New York Academy of Sciences, New York.

Guthrie, W.K.C., 1978, *A History of Greek Philosophy: Vol. II, The Presocratic Tradition from Parmenides to Democritus*, Cambridge University Press, Cambridge.

Haldane, J. B. S., 1932, *The Causes of Evolution*, Longmans, Green, London.

Harding, S., 1986, *The Feminist Question in Science*, Open University Press, Milton Keynes.

Hardy, A., 1965, *The Living Stream*, Collins, London.

Hare, R.M., 1952, *The Language of Morals*, Oxford University Press, Oxford.

Hayek, F., 1979, *The Counter-Revolution of Science*, LibertyPress, Indianapolis.

Hicks, J., 1985, *Evil and the God of Love*, Macmillan, London.

Hodgson, D., 2005, 'A Plain Person's Free Will', *Journal of Consciousness Studies,* 12, no. 1, pp. 53-58.

Holdgate, M., 1996, *From Care to Action*, Earthscan, London.

Hooykaas, R., 1977, *Religion and the Rise of Modern Science*, Scottish Academic Press, Edinburgh.

Hume, D., 1959, *A Treatise of Human Nature*, vol. 1, J. M. Dent, London (first published 1738).

Huxley, A., 1980, *The Human Situation*, St. Albans, Triad/Panther Books,

Huxley, J. S., 1942, *Evolution: The modern synthesis*, George Allen and Unwin, London (2nd ed., 1963).

Jackson, F., 1982, 'Epiphenomenal Qualia', *Philosophical Quarterly* 32, 127-136.

_____, 1986, 'What Mary didn't Know', *Journal of Philosophy* 83, 291-295.

Kane, R., 1998, *The Significance of Free Will*, Oxford University Press, Oxford.

_____, ed., 2005, *A Contemporary Introduction to Free Will*, Oxford University Press, Oxford.

Koertge, N., 1998, ed., *A House Built on Sand: Exposing Postmodernist Myths About Science*, Oxford University Press, Oxford.

Kripke, S., 1981, *Naming and Necessity*, Blackwell, Oxford.

Laing, R.D., 1965, *The Divided Self* (Harmondsworth, Penguin, Harmondsworth.

Langley, L., 2005, *Soldiers in the Laboratory*, Scientists for Global Responsibility, Folkstone.

Larrimore, M., 2001, *The Problem of Evil*, Blackwell, Oxford.

Latour, B., 1987, *Science in Action*, Open University Press, Milton Keynes.

Lenski, G. et al., 1995, *Human Societies: An Introduction to Macrosociology*, McGraw-Hill, New York.

Libet, B. 1985, 'Unconscious cerebral initiative and the role of conscious will in voluntary action', *Behavioral and Brain Sciences*, 8, pp. 529-566.

Little, M., 1994, 'Moral Realism', *Philosophical Books*, 35, pp. 145-153 and 225-233.

Lloyd Morgan, C., 1896, 'On Modification and Variation', *Science*, 4, pp. 733-40.

Locke, J., 1961, *An Essay Concerning Human Understanding*, Everyman, London

Macionis, J. and K. Plummer, 1997, *Sociology: A Global Introduction*, Prentice Hall, New York.

Mackie, J.L., 1977, *Ethics: Inventing Right and Wrong*, Penguin Books, Harmondsworth.

Marcuse, H., 1964, *One Dimensional Man*, Beacon Press, Boston.

Matthews, M.R., 1989, *The Scientific Background to Modern Philosophy*, Hackett, Indianapolis

Maxwell, N., 1966, 'Physics and Common Sense', *British Journal for the Philosophy of Science* 16, pp. 295-311.

_____, 1968a, 'Can there be Necessary Connections between Successive Events?', *British Journal for the Philosophy of Science*, vol. 19, pp. 1-25.

_____, 1968b, 'Understanding Sensations', *Australasian Journal of Philosophy*, vol. 46, pp. 127-146.

_____, 1972, 'A New Look at the Quantum Mechanical Problem of Measurement', *American Journal of Physics* 40, pp. 1431-5.

_____, 1973, 'The Problem of Measurement - Real or Imaginary?', *American Journal of Physics* 41, pp. 1022-5.

_____, 1974, 'The Rationality of Scientific Discovery', *Philosophy of Science* 41, pp. 123-153 and 247-295.

_____, 1976a, *What's Wrong With Science?*, Bran's Head Books, Frome (2nd edition, 2009, Pentire Press, London).

_____, 1976b, 'Towards a Micro Realistic Version of Quantum Mechanics, Parts I and II', *Foundations of Physics* 6, pp. 275-92 and 661-76.

_____, 1980, 'Science, Reason, Knowledge and Wisdom: A Critique of Specialism', *Inquiry 23*, pp. 19-81.

_____, 1982, 'Instead of Particles and Fields', *Foundations of Physics 12*, pp. 607-31.

_____, 1984, *From Knowledge to Wisdom*, Basil Blackwell, Oxford, 1984; 2nd edition Pentire Press, London, 2007.

_____, 1985, 'Methodological Problems of Neuroscience', in *Models of the Visual Cortex*, edited by D. Rose and V.G. Dobson, John Wiley and Sons, Chichester, pp. 11-21.

_____, 1988, 'Quantum Propensiton Theory: A Testable Resolution of the Wave/Particle Dilemma', *British Journal for the Philosophy of Science*, 39, pp. 1-50.

_____, 1993a, 'Induction and Scientific Realism', *British Journal for the Philosophy of Science* 44, pp. 61-79, 81-101 and 275-305.

_____, 1993b, 'Beyond Fapp: Three Approaches to Improving Orthodox Quantum Theory and An Experimental Test', in *Bell's Theorem and the Foundations of Modern Physics*, edited by A. van der Merwe, F. Selleri and G. Tarozzi, World Scientific, pp. 362-70.

_____, 1994, Particle Creation as the Quantum Condition for Probabilistic Events to Occur, *Physics Letters A 187*, pp. 351-355.

_____, 1995, 'A Philosopher Struggles to Understand Quantum Theory: Particle Creation and Wavepacket Reduction', in *Fundamental Problems in Quantum Physics*, edited by M. Ferrero and A. van der Merwe, Kluwer Academic, London, pp. 205-14.

_____, 1998, *The Comprehensibility of the Universe: A New Conception of Science*, Oxford University Press, Oxford (paperback ed., 2003).

_____, 2000, 'The Mind-Body Problem and Explanatory Dualism', *Philosophy*, vol., 75, pp. 49-71.

_____, 2001, *The Human World in the Physical Universe: Consciousness, Free Will and Evolution*, Rowman and Littlefield, Lanham, Maryland.

_____, 2002a, 'Cutting God in Half', *Philosophy Now*, vol. 35, March/April, pp. 22-25.

_____, 2002b, 'The Need for a Revolution in the Philosophy of Science', *Journal for General Philosophy of Science* 33, pp. 381-408.

_____, 2004a, *Is Science Neurotic?*, Imperial College Press, London.

_____, 2004b, 'Does Probabilism Solve the Great Quantum Mystery?', *Theoria* vol. 19/3, no. 51, pp. 321-336.

_____, 2005a, 'Popper, Kuhn, Lakatos and Aim-Oriented Empiricism', *Philosophia* 32, nos. 1-4, pp. 181-239.

_____, 2005b, 'Science versus Realization of Value, Not Determinism versus Choice', *Journal of Consciousness Studies* vol. 12, no. 1, pp. 53-58.

_____, 2006a, 'Learning to Live a Life of Value' in J. Merchey, ed., 2006, *Living a Life of Value*, Values of the Wise Press, pp. 383-95.

_____, 2006b, 'Three Problems about Consciousness and their Possible Resolution', PMS WIPS 005, Nov 15, http://www.petemandik.com/blog/?s=maxwell.

_____, 2007a, *From Knowledge to Wisdom: A Revolution for Science and the Humanities*, Pentire Press, London [2nd edition of Maxwell (1984), revised and extended].

_____, 2007b, 'The Disastrous War against Terrorism: Violence versus Enlightenment', Chapter 3 of *Terrorism Issues: Threat, Assessment, Consequences and Prevention*, ed. A. W. Merkidze, Nova Science Publishers, New York, 2007, pp. 111-133 (available on my website www.nick-maxwell.demon.co.uk).

_____, 2009, 'Are Universities Undergoing an Intellectual Revolution?', *Oxford Magazine*, No. 290, Eighth Week, Trinity Term, June, pp. 13-16.

_____, 2010a, 'A Priori Conjectural Knowledge in Physics', in *What Place for the A Priori*, edited by Michael Shaffer and Michael Veber, Open Court, Chicago.

_____, 2010b, 'Is the Quantum World Composed of Propensitons?', in *Probabilities, Causes and Propensities in Physics,* edited by Mauricio Suárez, Synthese Library, Springer, Dordrecht.

_____, 2010c, 'Popper's Paradoxical Pursuit of Natural Philosophy', in *Cambridge Companion to Popper*, edited by Jeremy Shearmur and Geoffrey Stokes, Cambridge University Press, Cambridge.

Mayr, E. 1963, *Animal Species and Evolution*, Oxford University Press, Oxford.

McCord, M., 1999, *Horrendous Evils and the Goodness of God*, Ithaca, N.Y.: Cornell University Press, Ithaca, N.Y.

McDowell, J., 1998, *Mind, Value and Reality*, Harvard University Press, Cambridge, Mass.

McGrath, A. and J.C. McGrath, 2007, *The Dawkins Delusion?*, Society for Promoting Christian Knowledge, London.

McHenry, L., ed., 2009, *Science and the Pursuit of Wisdom: Studies in the Philosophy of Nicholas Maxwell*, Ontos Verlag, Frankfurt.

McMullin, E., ed., 2005, *The Church and Galileo*, University of Notre Dame Press, Notre Dame, IN.

Miller, G., 2001, *The Mating Mind*, Vintage, London.

Monot, J., 1974, *Chance and Necessity*, Fontana, Glasgow.

Moore, G.E., 1903, *Principia Ethica*, Cambridge University Press, Cambridge.

Morgan, E., 1990, *The Scars of Evolution*, Penguin Books, London.

Morton, A., 2009, *Eating the Sun*, Fourth Estate, London.

Nagel, T., 1974, 'What Is It Like to Be a Bat?', *The Philosophical Review* 83, pp. 435-50.

Newton, I., 1962, *Principia*, University of California Press, Berkeley.

_____, 1932, *Opticks*, Dover Publications, New York (reprinting of the 4th edition of Newton's *Opticks* published in 1730).

Nietzsche, F., 2006, *The Gay Science*, Dover Publications, New York

Nowell-Smith, P., 1952, *Ethics*, Penguin Books, Harmondsworth.

Osborn, H.F., 1896, 'Ontogenic and Phylogenic Variation', *Science*, 4, pp. 786-9.

Penrose, R., 2004, *The Road to Reality*, Jonathan Cape, London.

Pirsig, R., 1974, *Zen and the Art of Motorcycle Maintenance*, Bodley Head, London.

Popper, K., 1959, *The Logic of Scientific Discovery*, Hutchinson, London.

_____, 1962, *The Poverty of Historicism*, Routledge and Kegan Paul, London.

_____, 1963, *Conjectures and Refutations*, Routledge and Kegan Paul, London.

_____, 1969, *The Open Society and Its Enemies*, Routledge and Kegan Paul, London.

_____, 1972, *Objective Knowledge*, Oxford University Press, Oxford.

_____, 1976, *Unended Quest*, Fontana, Glasgow.

_____, 1994, *The Myth of the Framework*, Routledge, London.

Redoni, P., 1988, *Galileo Heretic*, Allen Lane, London.

Rees, M., 2003, *Our Final Century*, Arrow Books, London.

Richardson, L., 2006, *What Terrorists Want*, John Murray, London.

Robb, D., and J. Heil, 2009, 'Mental Causation', *The Stanford Encyclopedia of Philosophy (Summer 2009 Edition)*, Edward N. Zalta (ed.), http://plato.stanford.edu/archives/sum2009/entries/mental-causation/.

Rogers, P. F., 2006, 'Peace Studies' in A. Collins, ed., *Contemporary Security Studies*, Oxford University Press, Oxford, Ch. 3.

Roszak, T., 1973, *Where the Wasteland Ends*, Faber and Faber, London.

Russell, E. S., 1916, *Form and Function: A Contribution to the History of Animal Morphology*, John Murray, London.

Ryle, G., 1949, *The Concept of Mind*, Hutchinson, London.

Schrödinger, E., 1958, *Mind and Matter*, Cambridge University Press, Cambridge.

Schwartz, B., 1987, *The Battle for Human Nature*, W. W. Norton, New York.

Shea, W. and M. Arigas, 2003, *Galileo in Rome: The Rise and Fall of a Troublesome Genius*, Oxford University Press, Oxford

Simpson, G. G., 1953, 'The Baldwin Effect, *Evolution* 7, pp. 110-117.

Smart, J.J.C., 1963, *Philosophy and Scientific Realism*, Routledge and Kegan Paul, London

Smith, D., 2003, *The Atlas of War and Peace*, Earthscan, London.

Snow, C. P., (1964) *The Two Cultures and a Second Look*, Cambridge University Press, Cambridge.

Sokal, A. and J. Bricmont, 1998, *Intellectual Impostures*, Profile Books, London.

Sternberg, R. J., ed., 1990, *Wisdom: Its Nature Origins and Development*, Cambridge University Press, Cambridge.

Sternberg, R. J. et al., 2007, 'Teaching for wisdom: what matters is not just what students know, but how they use it', *London Review of Education*, 5, pp. 143-158.

Stevenson, C.L., 1944, *Ethics and Language*, Yale University Press, New Haven, Conn.

Swinburne, R., ed., 1974, *The Justification of Induction*, Oxford University Press, Oxford

_____, 2003, *Providence and the Problem of Evil*, Oxford Scholarship Online.

Tischler, H., 1996, *Introduction to Sociology*, Harcourt Brace, Orlando.

Trivers, R., 1985, *Social Evolution*, Benjamin/Cummings Publishing, Menlo Park.

Troyat, H., 1970, *Tolstoy*, Penguin Books, London

337

Tyndall Centre, ed., 2006, *Truly Useful*, Tyndall Centre, UK.

van Inwagen, P., 1986, *An Essay on Free Will*, Clarendon Press, Oxford.

_____, 2006, *The Problem of Evil*, Oxford University Press, Oxford.

Waddington, C. H., 1957, *The Strategy of the Genes*, Allen and Unwin, London.

Walker, G. and D. King, 2008, *The Hot Topic*, Bloomsbury, London.

Wallace, D., 2008, 'The Philosophy of Quantum Theory', in *Ashgate Companion to the New Philosophy of Physics*, edited by D. Rickles, Ashgate, London.

Weart, S., 2003, *The Discovery of Global Warming*, Harvard University Press, Cambridge, Mass.

Weber, B. and D. Depew, eds., 2003, *Evolution and Learning: The Baldwin Effect Reconsidered*, MIT Press, Cambridge, Mass.

White, M., 2006, *The Pope and the Heretic*, Abacus, London.

White, M., 2007, *Galileo: Antichrist: A Biography*, Weidenfeld & Nicolson, London.

Wilsdon, J. and R. Willis, 2004, *See-through Science*, Demos, London.

Zeki, S., 1993, *A Vision of the Brain*, Blackwell Scientific Publications, London.

INDEX

of science 141-7, 176-7
d'Alembert, J. 45, 170n
Allen, W. 99
Anaximander 52n
Appleyard, B. 194n
aquatic ape hypothesis 288
Arigas, M. 23n
Aristotle 23, 52n, 126, 207n
armaments 43, 302, 309
  and science 47, 165, 304
Ayer, A.J. 104n

Bach, J.S. 111
Bacon, F. 43, 168, 170n
Baldwin, M. 285-8
Baldwin effect 285-8
Barnes, B. 195n
Bartley, B. 14
Bateson, P. 286n, 289
Beethoven, L. van 29, 89
behaviourism 81, 230n
Bell, J.S. 162n
Berkeley, G. 75n
Berlin, I. 29n, 169, 194n
Berman, M. 194n
Bernanos, G. 7n
bin Laden, A. 49
Blair, T. 49
Blake, W. 29, 119
Bloor, D, 195n
Boesiger, E. 281
Bohm, D. 162
Bohr, N. 157
Bond, E.J. 104n
Bonner, J. 282, 288
Born, M. 157, 163
Bowler, P. 288
Boyd, R. 280, 281, 285n, 288
Boyle, R. 24, 44
Bricmont, J. 169n, 195
Brink, D. 104n

Broglie, L. de 159, 162
Brown, G. 209-10, 212, 215,
  217-8, 244, 246
Bruno, G. 23, 24, 44
Bush, G. 49
Campbell, D. 281n
Cartesian dualism 63-5, 121,
  202, 204
causation 150-1, 304-5
  ambiguities of 304-5
  anti-Humean account of 150-5
Christianity 23-6, 100-1
  and persecution 23-4
  and science 24-6
Churchland, P. 251n
Coleridge, S.T. 29
Compatibilism, definitions of
  201n-202n, 208n
Compatibilist/Incompatibilist
  debate 208-45, 264n
  see also free will, problem of
comprehensibility of the
  universe 124, 127-9, 131, 142-
  4, 171-2
Condorcet, N. de 25, 167
conjectural essentialism 152-5
conjectural objectivism 106
consciousness 77-82, 228-44
  and science 77-8
  location in brain 223n
  mystery of solved 77-82
control aspect of brains 81, 219,
  221-33
  knowledge of 221-4
control systems 205-7, 212-3
Copernicus 24
Cronin, H. 269n, 285n
cutting God in half 8-9, 15
  problems created by 9, 15-18

Dampier, D. 25n
dark matter 135n-136n

340

Galileo 23, 24, 44, 60-1, 90,
126, 141, 207n
Gandhi, M. K. 19
Gascardi, A. 169n
Gatti, H, 23n
Gay, P. 25n, 45n, 167n, 168n
Ghirardi, G.C. 162
Ginzburg, C. 23n
global problems 18, 20-2, 165-
6, 298-304
    and knowledge-inquiry 315-7
    and wisdom-inquiry 312-5,
    317
    caused by science 166, 303-5
    science not the problem 305-6
    what needs to be done 307-12
global warming 21, 47, 165,
299-300, 326
    and wisdom-inquiry 299-300
    what needs to be done 307-8
God 1-14
    and education 1, 5
    and problem of evil 3-5
    and the Devil 5, 7-8
    broad conception of 6-14
    Christian 1, 9
    conjectural character of 2
    death of 12
    doctrine of 1
    Einstein's 1, 8
    evil of 3-6
    excuses for 4-5
    how to characterize 6-14
    Islamic 1, 9
    Judaic 1, 9
    metaphorical version of 13-14
    need to cut in half 1, 8-9
    need to improve conception of
    12-13
    problematic character of 3-8
    problem-solving power of 2-3,

12
    putting together 1, 96
    refutation of existence of 3-5
    significance of 1-3, 12-13
    traditional attributes of 1
    see also God-of-Cosmic-
    Power, God-of-Cosmic-Value
God-of-Cosmic-Power 1, 8-9,
15, 122-63
    and aim-oriented empiricism
    123-4, 129, 131-41, 142-50,
    158-160, 163
    and conjectural essentialism
    152-5
    and explanation 123, 131,
    154-5
    and falsificationism 130-1
    and God-of-Cosmic-Value 53,
    54, 96, 200
    and ideas about how the
    universe is physically
    comprehensible 125, 135-55
    and metaphysical assumptions
    of science 123-8, 130-4, 135-
    55
    and necessary connections
    between successive events
    151-5
    and necessitating properties
    151-5
    and new conception of science
    123-4, 126-7, 131-50
    and persons 203, 208
    and physical entities 125
    and physicalism 143, 147
    and physical theory 126-8,
    131-47
    and physics 122-63
    and quantum theory 122, 126,
    127, 131n, 135, 144, 155-63
    and rigour 148-9
    and science 122-63

344

and scientific revolutions 126,
144
and standard empiricism 122-
3, 128-9, 132, 142, 147, 148-9
and string theory 123, 125
and theoretical unity 123-4,
131-4, 135-41
and traditional conceptions of
God 150-1, 154
and unified theory of
everything 123
and unifying power of physics
126-8, 131-4
argument for unknown
character of 122-3, 125-6
conjectural character of 130-1
existence established by
science 122-4, 127-8, 131-4
questions concerning 128
unscientific character of 122-3
we are a part of 15, 203, 208,
253, 264
what it is 122, 135-41, 143-4,
199
why it can be forgiven 8-9,
150
God-of-Cosmic-Value 9, 15, 84-
121
and choice 92, 101-2, 103
and concern for others 19, 90,
119
and conjectural objectivism
106, 109-117
and control 104-5
and cooperation 19, 90
and conscious life 85
and criticism 86-7, 91, 92-5
and death 90-1, 98-102
and desires and feelings 86-7
and despair 97-8
and dogmatic objectivism
106-9, 116

and environment 86
and epistemology 106, 114-6
and equality 102-3
and fallibilism 91, 92, 96, 97-
8, 109-117
and friendship 18-19, 90, 119
and God-of-Cosmic-Power
53, 54, 96
and higher spiritual values
119-121
and hypocrisy 96, 104
and interests 117-8
and liberalism 105, 107, 109-
110, 114
and love 18-19, 90, 92-6, 119
and metaphysics 105-6, 110-
114
and morality 104-5
and mystical experience 119-
121
and persons 85-6, 117-8
and realism 86-7, 88, 104-116
and reason 103
and relativism 104-116
and selfishness 18-19
and sentient life 85
and subjectivity 104-116
and suffering 90-1
and tragedy 88
and wisdom-inquiry 34, 35-8,
43, 51, 119, 121, 328
diverse character of 87
evolutionary character of 89-
90
existence of 97-8, 104-117
in us 10, 15, 84
moral implications of 18-19,
90, 92-6, 104
origin of 88-9
refutation of bad arguments
for subjectivity of 104-116
what it is 9, 15, 84-121

God-thesis 1-14
  need to take seriously 1-2
  problem-solving power of 2-3
  problems of 3-8
  refutation of 3-5
  taken seriously 2-13
Goethe, J. W. von 29
Goodall, J. 119
Grice, P. 293
Gross, P. 169n
Guthrie, W.K.C. 60n

Habermas, J. 169n
Haldane, J.B.S. 284
Harding, S. 196n
Hardy, A. 284, 286n, 287-9
Hare, R.M. 104n
Hayek, F. 45n
Hazlitt, W. 29
head processes 79-82, 204
  control aspects of 81
  mental aspects of 74-82
  theory 81-2
  two aspects of 79-82, 204
Heil, J. 220n
Heisenberg, W. 157
Heraclitus 52n
Hicks, J. 4n
Hitler, A. 19, 102
Hodgson, D. 215n, 249n
Holdgate, M. 166n
Hooke, R. 24, 44
Hooykaas, R. 24n
Horkheimer, M. 169n
human suffering 2-3, 5-6, 8-9,
  20-1, 24, 43-4, 90-1
human world/physical universe
  problem 9-11, 15-16, 53-83
  religious character of 9-11
  significance of 15-16
humanism 9, 12

religious character of 9
Hume, D. 150-1, 153
Hutton, J. 288
Huxley, A. 165n
Huxley, J. 284, 287
Huxley, T. H. 25
  and agnosticism 25
Huygens, C. 24, 44

incompatibilism 202n
internalism 65-76, 202, 204, 233
  arguments for refuted 66-70
  arguments against 74-7
  definition of 65-6

Jackson, F. 56
Jones, S. 300

Kane, R. 202n, 215n
Kant, I. 170n
Keats, J. 29, 119
Kepler, J. 24, 90
King, D. 48n
knowledge-inquiry 26-9
  and empiricism 42
  and global problems 315-7
  and problems 35
  and science 28-9, 38
  and social inquiry 38-9
  and society 39-40
  censorship system of 27-8
  content of 41
  damaging effects of 32-4, 43,
  186-8
  history of 23-6,
  how differs from wisdom-
  inquiry 35-43, 184-8
  influence of 27, 31-2n, 320-1
  irrationality of 27, 29, 30, 31,
  45, 51
  see also academia, aim-
  oriented empiricism,

and empathic understanding
247
and folk psychology 247
and free will 248-9
and knowledge-inquiry 251-2
and natural science 253
and non-reducibility to
physics 249-51
and purposive explanation 248
and theory of mind 247
and wisdom-inquiry 251-2
authenticity of 251-4
personalistic understanding 188
persons
and physics 200-2
existence of 199-202
philosophy
a new approach? 51-2
analytic 51
basic task of 51-2
Continental 51
fundamental problem of 10,
16
of loving 90, 92-6
physical theory
and aim-oriented empiricism
129, 131-4, 143-6
and falsificationism 130
and solution to problem of
verisimilitude 149-50
essentialistic interpretation of
151-5
of everything 8, 15, 57-8, 60,
122, 132, 136, 140, 143-5,
147, 154-5, 203
revolutionary development of
126-7, 144
unity of 123-4, 126-8, 131-41
see also aim-oriented
empiricism, Einstein, God-of-
Cosmic-Power, Maxwell,
Newton, physics, quantum

theory, standard empiricism,
unity of theory
physical world 53-6
aspect of things 54-9
two aspects of 136-7
physicalism 143, 147, 201, 203,
206, 208, 209-211
definition of 143, 201n
probabilistic 210, 214-6
physics
and causally efficacious 54,
55, 59, 73
and experiential 53-65, 73, 74,
199
annihilates experiential world
53-4, 76-7
aims of 171-2
comprehensive character of
58-9
explanation for restricted
character of 58-9, 73, 199
limitations of 56n, 57n
orthodox view of 59-63
religious character of 11, 124
Pirsig, R. 105
Plato 4n, 23, 52n
politics 16
religious character of 11
Popper, K. 2, 17n, 30n, 97,
102n, 123n, 129, 284
and critical rationalism 173-4,
191
and evolution 284
and scientific method 130,
173-4
and specialization 30n, 173n-
4n
and social engineering 192-4
and standard empiricism 170n
population growth 21, 33, 43,
47, 165, 179, 302
what needs to be done 310

Richardson, L. 50n
Richerson, P. 280, 281, 285n, 288
Rimini, A. 162
Robb, D. 220n
Rogers, P.F. 325n
Romanticism 29, 169
Rorty, R. 169n
Roszak, T. 194n
Rousseau, J.-J. 29
Russell, E.S. 287
Ryle, G. 192

Scandella, D. 23, 44
Schiller, F. 29
Schrödinger, E. 159, 163, 287-8
Schubert, F. 29
Schwartz, B. 194n
science 16, 20-21
  and Christianity 24-25, 44
  and global problems 20-21
  and social life 195-7
  danger of 20-21, 165-6
  empirical character of 28-9
  fundamental problem of 9
  new conception of 123-4, 126-7, 131-50
  problematic aims of 141-2, 176-7
  religious character of 9, 124
  success of 20-21, 164
  see also aim-oriented empiricism, knowledge-inquiry, wisdom-inquiry
scientific aim
  of enhancing quality of human life 177
  of explanatory truth 171-2, 176
  of truth 170-1, 175

of valuable truth 176-7
scientific method
  aim-oriented empiricist conception of 129, 142-7, 171-2
  denial of 195
  generalizing 166-9, 172-5
  Popper's view of 130, 173-4
  standard empiricist conception of 129, 170
  see also aim-oriented empiricism, standard empiricism
scientific revolutions 125-7, 144
sensation/brain process
  correlations 237-9
  explanation for 237-9
sensory qualities 79
  objective existence of 71-4
Shakespeare, W. 89, 99
Shea, W. 23n
Simpson, G.G, 285-7
Smart, J.J.C. 66n, 68n
social inquiry 38-9
social science 25-6, 38-9, 46
Socrates 52n
Sokal, A. 169n, 195n
Snow, C.P. 183n
specialization 32, 34-5
  and Popper 30n, 173n-4n
Spinoza, B. 56
Stalin 4, 5, 19, 102
standard empiricism 122-3, 128-9, 132
  and Newton 171n
  formulation of 129, 170
  lack of rigour 148-9
  Popper's formulation of 170n
  refutation of 131-4, 170-2
  see also knowledge-inquiry
standard model 127, 135, 144, 199

350

Stendhal 4n
Sternberg, R. 326
Stevenson, C.L. 104n
string theory 123
subjectivity 71-2
  two senses of 71-2
summary of book 16-18, 328
Swinburne, R. 4n, 54n

technology 20-21
  and global problems 20-21,
  165-6
  and war 21
  success of 20-21
  danger of 20-21
terrorism, war against 48-50
Thales 52n
theodicy 4n
Tolstoy, L. 82
Trivers, R. 282, 288
Troyat, H. 82n
Truffaut, F. 289n
two-aspect view 56-9

United Nations 308-9, 311
unity in nature 123-4, 126-8,
  134
  and common sense 134
unity of theory 123-4, 131-41
  and explanation 123, 131, 140
  and simplicity 140n
  and symmetry 139n
  concerns content, not form
  136, 139
  degrees of 135-41
  kinds of 137-9
  requirement for 140
  what it is 135-41

value 84-121
  and choice 92, 101-2, 103
  and concern for others 19, 90,

119
and conjectural objectivism
106, 109-117
and conscious life 85
and control 104-5
and cooperation 19, 90
and criticism 86-7, 91, 92-5
and death 90-1, 98-102
and desires and feelings 86-7
and despair 97-8
and dogmatic objectivism
106-9, 116
and epistemology 106, 114-6
and environment 86
and equality 102-3
and fallibilism 91, 92, 96, 97-
8, 109-117
and friendship 18-19, 90, 119
and God-of-Cosmic-Power
53, 54, 96
and hypocrisy 96, 104
and interests 117-8
and liberalism 105, 107, 109-
110, 114
and love 18-19, 90, 92-6, 119
and metaphysics 105-6, 110-
114
and morality 104-5
and mystical experience 119-
121
and persons 85-6, 117-8
and realism 86-7, 88, 104-116
and reason 103
and relativism 104-116
and selfishness 18-19
and sentient life 85
and subjectivity 104-116
and suffering 90-1
and tragedy 88
and wisdom-inquiry 34, 35-8,
43, 51, 119, 121
diversity of 87

351

evolutionary character of 89-90
existence of 97-8, 104-117
higher spiritual 119-121
ideas about 17, 84-121
in us 10, 15, 84
intrinsic 85
moral implications of 18-19, 90, 92-6, 104
origin of 88-9
poles of 90-1
properties 105-6, 110-114
questions concerning 84-5
refutation of bad arguments for subjectivity of 104-116
what is of 9, 15, 84-121
see also God-of-Cosmic-Value, wisdom, wisdom-inquiry
van Inwagen, P. 4n, 213n
verisimilitude, problem of 149-50
vitalism 216
Voltaire, F. 25, 45, 167

Waddington, C.H. 287, 289
Walker, G. 48n
Wallace, D. 162n
war 1, 20-1, 301
  what needs to be done 308-9
Weart, S. 48n
Weber, B. 285, 288
Weber, M. 168n
Weber, T. 162
White, M. 23n
will power 257-61
  and Christianity 257-9
  and conflict resolution 260-1
  and Darwinism 260-1
Willis, R. 325n
Wilsdon, J. 325n
wisdom 17n, 265

definition of 17n, 182, 254n,
multi-dimensional character of 255-6
presupposes free will 254-7
wisdom-inquiry 17, 27, 34-43, 163-98
  and aim-oriented empiricism 171-2, 174-7
  and aim-oriented rationalism 174-82
  and aims of science 170-7
  and art 183
  and civilization 179-80
  and Einstein 186
  and empiricism 42, 181
  and Enlightenment blunders 169-79
  and evolution 297-300
  and experience 42, 181
  and feelings and desires 182-3, 186
  and financial crisis 299
  and global problems 312-5, 317
  and global warming 47-8, 50, 299-300
  and google 326-7
  and humanities 178-81
  and institutions 178-82
  and knowledge versus wisdom 190-2
  and New Enlightenment 170-84
  objections to 317-20
  and personalistic understanding 188
  and problems of living 35-8, 181
  and rational improvement of philosophies of life 179-81
  and rules of reason 34-5
  and science 39, 164-77

COMMENTS ON WORK BY

NICHOLAS MAXWELL

From Knowledge to Wisdom (1984, Blackwell; 2nd ed., 2007, Pentire Press)

"The essential idea is really so simple, so transparently right ... It is a profound book, refreshingly unpretentious, and deserves to be read, refined and implemented."
Dr. Stewart Richards, *Annals of Science*

"Maxwell's book is a major contribution to current work on the intellectual status and social functions of science ... [It] comes as an enormous breath of fresh air, for here is a philosopher of science with enough backbone to offer root and branch criticism of scientific practices and to call for their reform."
Dr. David Collingridge, *Social Studies of Science*

"Maxwell has, I believe, written a very important book which will resonate in the years to come. For those who are not inextricably and cynically locked into the power and career structure of academia with its government-industrial-military connections, this is a book to read, think about, and act on."
Dr. Brian Easlea, *Journal of Applied Philosophy*

"In this book, Nicholas Maxwell argues powerfully for an intellectual "revolution" transforming all branches of science and technology. Unlike such revolutions as those described by Thomas Kuhn, which affect knowledge about some aspect of the physical world, Maxwell's revolution involves radical changes in the aims, methods, and products of scientific inquiry, changes that will give priority to the personal and social problems that people face in their efforts to achieve what is valuable and desirable."
Professor George Kneller, *Canadian Journal of Education*

"Any philosopher or other person who seeks wisdom should read this book. Any educator who loves education – especially those in leadership positions – should read this book. Anyone who wants to understand an important source of modern human malaise should read this book. And anyone trying to figure out why, in a world that produces so many technical wonders, there is such an immense "wisdom gap" should read this book. In *From Knowledge to Wisdom: A Revolution for Science and the Humanities....* Nicholas Maxwell presents a compelling, wise, humane, and timely argument for a shift in our fundamental "aim of inquiry" from that of knowledge to that of wisdom."
Jeff Huggins *Metapsychology*

"This book is the work of an unashamed idealist; but it is none the worse for that. The author is a philosopher of science who holds the plain man's view that philosophy should be a guide to life, not just a cure for intellectual headaches. He believes, and argues with passion and conviction, that the abysmal failure of science to free society from poverty, hunger and fear is due to a fatal flaw in the accepted aim of scientific endeavour – the acquisition and extension of knowledge. It is impossible to do Maxwell's argument justice in a few sentences, but, essentially, it is this. At the present time the pursuit of science – indeed the whole of academic inquiry – is largely dominated by 'the philosophy of knowledge'. At the heart of this philosophy is the assumption that knowledge is to be pursued for its own sake. But the pursuit of objective truth must not be distorted by human wishes and desires, so scientific research becomes divorced from human needs, and a well-intentioned impartiality gives way to a deplorable indifference to the human condition. The only escape is to reformulate the goals of science within a 'philosophy of wisdom', which puts human life first and gives 'absolute priority to the intellectual tasks of articulating our problems of living, proposing and criticizing possible solutions, possible and actual human actions'. The philosophy of wisdom commends itself, furthermore, not only to the heart but to the head: it gives science and scholarship a proper place in the human social order. . . Nicholas Maxwell has breached the conventions of philosophical writing by using, with intent, such

355

loaded words as 'wisdom', 'suffering' and 'love'. 'That which is of value in existence, associated with human life, is inconceivably, unimaginably, richly diverse in character.' What an un-academic proposition to flow from the pen of a lecturer in the philosophy of science; but what a condemnation of the academic outlook, that this should be so

Professor Christopher Longuet-Higgins, *Nature*

"Wisdom, as Maxwell's own experience shows, has been outlawed from the western academic and intellectual system ... In such a climate, Maxwell's effort to get a hearing on behalf of wisdom is indeed praiseworthy."

Dr. Ziauddin Sardar, *Inquiry*

"This book is a provocative and sustained argument for a 'revolution', a call for a 'sweeping, holistic change in the overall aims and methods of institutionalized inquiry and education, from knowledge to wisdom' ... Maxwell offers solid and convincing arguments for the exciting and important thesis that rational research and debate among professionals concerning values and their realization is both possible and ought to be undertaken."

Professor Jeff Foss, *Canadian Philosophical Review*

"Maxwell's argument ... is a powerful one. His critique of the underlying empiricism of the philosophy of knowledge is coherent and well argued, as is his defence of the philosophy of wisdom. Most interesting, perhaps, from a philosophical viewpoint, is his analysis of the social and human sciences and the humanities, which have always posed problems to more orthodox philosophers, wishing to reconcile them with the natural sciences. In Maxwell's schema they pose no such problems, featuring primarily ... as methodologies, aiding our pursuit of our diverse social and personal endeavours. This is an exciting and important work, which should be read by all students of the philosophy of science. It also provides a framework for historical analysis and should be of interest to all but the most blinkered of historians of science and philosophy."

Dr. John Hendry, *British Journal for the History of Science*

"… a major source of priorities, funds and graduates' jobs in 'pure science' is military … this aspect of science is deemed irrelevant by the overwhelming majority of those who research, teach, sociologize, philosophise or moralize about science. What are we to make of such a phenomenon? It is in part a political situation, in its causes and effects; but it is also philosophical, and this is Nick Maxwell's point of focus. Such a gigantic co-operative endeavour of concealment, amounting to a huge deception, could be accomplished naturally by all educated, humane participants, a 'conspiracy needing no conspirators', only because their 'philosophy of knowledge' envelops them in the assurance that their directors, paymasters and employers have nothing to do with the real thing – the research. This, to me, is the heart of Maxwell's message."
Dr. Jerry Ravetz, *British Journal for the Philosophy of Science*

"This book is written in simple straightforward language … The style is passionate, committed, serious; it communicates Maxwell's conviction that we are in deep trouble, that there is a remedy available, and that it is ingrained bad intellectual habits that prevent us from improving our lot … Maxwell is raising an important and fundamental question and things are not going so well for us that we should afford the luxury of listening only to well-tempered answers."
Professor John Kekes, *Inquiry*

"Because Maxwell so obviously understands and loves science as practiced, say, by an Einstein, his criticisms of current science seem to arise out of a sadness at missed opportunities rather than hostility … I found Maxwell's exposition and critique of the current state of establishment science to be clear and convincing … Maxwell is right to remind us that in an age of Star Wars and impending ecological disaster, talk of the positive potential of means-oriented science can easily become an escapist fantasy."
Professor Noretta Koertge, *Isis*

"In an admirable book called *From Knowledge to Wisdom*, Nicholas Maxwell has argued that the radical, wasteful

misdirection of our whole academic effort is actually a central cause of the sorrows and dangers of our age . . .Thinking out how to live is a more basic and urgent use of the human intellect than the discovery of any fact whatsoever, and the considerations it reveals ought to guide us in our search for knowledge. . . In arguing this point . . . Maxwell proposes that we should replace the notion of aiming at knowledge by that of aiming at wisdom. I think this is basically the right proposal. . . Maxwell is surely right in saying that [the distorted pursuit of knowledge], because it wastes our intellectual powers, has played a serious part in distorting our lives."
Mary Midgley, *Wisdom, Information and Wonder*

"[T]here is...much of interest and, yes, much of value in this book...Maxwell is one of those rare professional philosophers who sees a problem in the divorce between thought and life which has characterized much of modern philosophy (and on both sides of the English channel, not merely in the so-called 'analytic' tradition'); he wishes to see thought applied to life and used to improve it. As a result, many of the issues he raises are of the first importance ... He has . . produced a work which should give all philosophers and philosophically-minded scientists cause for reflection on their various endeavors; in particular, it should give philosophers who are content to be specialists a few sleepless nights."
Professor Steven Yates, *Metaphilosophy*

"Maxwell [argues for] an "intellectual revolution" that will affect the fundamental methods of inquiry of science, technology, scholarship and education, looking not for knowledge for knowledge's sake, but for wisdom, which he says is more rational and of greater human value and holds the potential to alleviate human problems and institute social change. A humanist and philosopher, Maxwell presents his ideas with eloquence and conviction. This book will appeal to persons in many different disciplines – from science to social studies."
American Library Association

"Maxwell's thesis is that the evident failure of science to free society from poverty, hunger and the threat of extinction results from a 'fatal flaw in the accepted aim of scientific endeavour'. . . It is precisely because of 'the accepted aim' that acquisition of knowledge, which presumably originated as an essential strategy for survival, has given rise to the relentless pursuit of new and better ways of achieving the exact opposite. . . For Maxwell, the solution is obvious – a radically new approach to the whole business of intellectual inquiry. . . It is hard to argue with these aims . . . If we could only change the way people feel, Maxwell's solution would be easier, if not easy."
Professor Norman F. Dixon, *Our Own Worst Enemy*

"a sustained piece of philosophical reasoning which makes a real contribution to the reinstatement of philosophy as a central concern. We need to follow Maxwell's lead in constructing a philosophy of wisdom."
P. Eichman, *Perspectives on Science and Christian Faith*

"Nicholas Maxwell (1984) defines freedom as 'the capacity to achieve what is of value in a range of circumstances'. I think this is about as good a short definition of freedom as could be. In particular, it appropriately leaves wide open the question of just what is of value. Our unique ability to reconsider our deepest convictions about what makes life worth living obliges us to take seriously the discovery that there is no palpable constraint on what we can consider."
Professor Daniel Dennett, *Freedom Evolving*

"The Rationality of Scientific Discovery", *Philosophy of Science* (1974)

"Maxwell's theory of aim-oriented empiricism is the outstanding work on scientific change since Lakatos, and his thesis is surely correct. Scientific growth should be rationally directed through the discussion, choice, and modification of aim-incorporating blueprints rather than left to haphazard competition among research traditions seeking empirical success alone. . . Of the

theories of scientific change and rationality that I know, Maxwell's is my first choice. It is broad in scope, closely and powerfully argued, and is in keeping with the purpose of this book, which is to see science in its totality. No other theory provides, as Maxwell's does in principle, for the rational direction of the overall growth of science."
Professor George F. Kneller, *Science as a Human Endeavor*

"As Nicholas Maxwell has suggested, if we make one crucial assumption about the purpose of science, then the possibility arises that some paradigms and theories can be evaluated even prior to the examination of their substantive products. This one crucial assumption is that the overall aim of science is to discover the maximum amount of order inherent in the universe or in any field of inquiry. Maxwell calls this 'aim-oriented empiricism'. . . I agree with Maxwell's evaluation of the importance of coherent aim-oriented paradigms as a criterion of science. . . The time is ripe, therefore, to replace the incoherent and unconscious paradigms under whose auspices most anthropologists conduct their research with explicit descriptions of basic objectives, rules, and assumptions. That is why I have written this book."
Professor Marvin Harris, *Cultural Materialism*

The Comprehensibility of the Universe: A New Conception of Science (1998, paperback 2003, Oxford University Press)

"Nicholas Maxwell's ambitious aim is to reform not only our philosophical understanding of science but the methodology of scientists themselves ... Maxwell's aim-oriented empiricism [is] intelligible and persuasive ... the main ideas are important and appealing ... an important contribution to the philosophy of physics."
J. J. C. Smart, *British Journal for the Philosophy of Science*

"Maxwell has clearly spent a lifetime thinking about these matters and passionately seeks a philosophical conception of science that will aid in the development of an intelligible physical worldview.

He has much of interest to say about the development of physical thought since Newton. His comprehensive coverage and sophisticated treatment of basic problems within the philosophy of science make the book well worth studying for philosophers of science as well as for scientists interested in philosophical and methodological matters pertaining to science."
Professor Cory F. Juhl, *International Philosophical Quarterly*

"Maxwell performs a heroic feat in making the physics accessible to the non-physicist ... Philosophically, there is much here to stimulate and provoke . . . there are rewarding comparisons to be made between the functional roles assigned to Maxwell's metaphysical "blueprints" and Thomas Kuhn's paradigms, as well as between Maxwell's description of theoretical development and Imre Lakatos's methodology of scientific research programmes."
Dr.Anjan Chakravartty, *Times Higher Education Supplement*

"some of [Maxwell's] insights are of everlasting importance to the philosophy of science, the fact that he stands on the shoulders of giants (Hume, Popper) notwithstanding . . . My overall conclusion is that Universe is an ideal book for a reading group in philosophy of science or in philosophy of physics. Many of the pressing problems of the philosophy of science are discussed in a lively manner, controversial solutions are passionately defended and some new insights are provided; in particular the chapter on simplicity in physics deserves to be read by all philosophers of physics."
Dr. F. A. Muller, *Studies in History and Philosophy of Modern Physics*

"Maxwell ... has shown that it is absurd to believe that science can proceed without some basic assumptions about the comprehensibility of the universe . . . Throughout this book, Maxwell has meticulously argued for the superiority of his view by providing detailed examples from the history of physics and mathematics . . . The Comprehensibility of the Universe attempts to resurrect an ideal of modern philosophy: to make rational sense of science by offering a philosophical program for improving our

361

knowledge and understanding of the universe. It is a consistent plea for articulating the metaphysical presuppositions of modern science and offers a cure for the theoretical schizophrenia resulting from acceptance of incoherent principles at the base of scientific theory."
Professor Leemon McHenry, *Mind*

"This admirably ambitious book contains more thought-provoking material than can even be mentioned here. Maxwell's treatment of the descriptive problem of simplicity, and his novel proposals about quantum mechanics deserve special note. In his view the simplicity of a theory is (and should be) judged by the degree to which it exemplifies the current blueprint of physicalism, that blueprint determining the terminology in which the theory and its rivals should be compared. This means that the simplicity of a theory amounts to the unity of its ontology, a view that allows Maxwell to offer an explanation of our conflicting intuitions that terminology matters to simplicity, and that it is utterly irrelevant. Maxwell's distinctive views about what is wrong with quantum mechanics grow out of his adherence to aim-oriented empiricism: the much-discussed problem of measurement is for him a superficial consequence of the deeper problem that the ontology of the theory is not unified, in that no one understands how one entity could be both a wave and a particle. In response to this problem Maxwell finds between the metaphysical cracks a way to fuse micro-realism and probabilism, which leads him to a proposal to solve the measurement problem by supplementing quantum mechanics with a collapse theory distinct from the recent and popular one of Ghirardi, Rimini and Weber. Maxwell's highly informed discussions of the changing ontologies of various modern physical theories are enjoyable, and the physical and mathematical appendix of the book should be a great help to the beginner."
Professor Sherrilyn Roush, *The Philosophical Review*

"Nicholas Maxwell has struck an excellent balance between science and philosophy . . . The detailed discussions of theoretical unification in physics - from Newton, Maxwell and Einstein to Feynman, Weinberg and Salam - form some of the best material

in the book. Maxwell is good at explaining physics . . . Through the interplay of metaphysical assumptions, at varying distances from the empirical evidence Maxwell shows, rather convincingly, that in the pursuit of rational science the inference from the evidence to a small number of acceptable theories, out of the pool of rival ones, is justifiable . . . Its greatest virtue is the detailed programme for a modern version of natural philosophy. Along the way, Maxwell homes in on the notion of comprehensibility by the exclusion of less attractive alternatives. In an age of excessive specialization the book offers a timely reminder of the close link between science and philosophy. There is a beautiful balance between concrete science and abstract philosophy . . . In the "excellently written Appendix some of the basic mathematical technicalities, including the principles of quantum mechanics, are very well explained . . . Einstein held that 'epistemology without science becomes an empty scheme' while 'science without epistemology is primitive and muddled'. Maxwell's new book is a long-running commentary on this aphorism."
Dr. Friedel Weinert, *Philosophy*

"In *The Comprehensibility of the Universe*, Nicholas Maxwell develops a bold, new conception of the relationship between philosophy and science...Maxwell has a metaphysically rich, evolutionary vision of the self-correcting nature of science...The work is important...An added benefit of Maxwell's analysis...is the possibility of a positive, fruitful relationship to emerge between science and the philosophy of science...his important and timely critique of the reigning empiricist orthodoxy...what does it mean to say simplicity is a theoretical virtue? And why should we prefer simple to complex theories? Maxwell provides an admirable discussion of these issues. He also provides a useful discussion of simplicity in the context of theory unification – simple theories are unifying theories – and illustrates his points with examples drawn from Newtonian physics and Maxwellian electrodynamics...It is hard to do justice to the richness of Maxwell's discussion in this chapter. I can only say that this is a chapter that will repay serious study...Maxwell turns his attention to issues surrounding the theoretical character of evidence, the idea of scientific progress and

the question as to whether there is a method of discovery....The discussion of these matters – as with the other topics covered in this book – is conceptually rich and technically sophisticated. A useful antidote, in fact, to the settled orthodoxy surrounding these philosophical issues...Maxwell has written a book that aims to put the metaphysics back in physics. It is ambitious in scope, well-argued, and deserves to be seriously studied."
Professor Niall Shanks, *Metascience*

The Human World in the Physical Universe: Consciousness, Free Will and Evolution (2001, Rowman and Littlefield)

"Ambitious and carefully-argued...I strongly recommend this book. It presents a version of compatibilism that attempts to do real justice to common sense ideas of free will, value, and meaning, and...it deals with many aspects of the most fundamental problems of existence."
Dr. David Hodgson, *Journal of Consciousness Studies*

"Maxwell has not only succeeded in bringing together the various different subjects that make up the human world/physical universe problem in a single volume, he has done so in a comprehensive, lucid and, above all, readable way."
Dr. M. Iredale, *Trends in Cognitive Sciences*

"...a bald summary of this interesting and passionately-argued book does insufficient justice to the subtlety of many of the detailed arguments it contains."
Professor Bernard Harrison, *Mind*

"Nicholas Maxwell takes on the ambitious project of explaining, both epistemologically and metaphysically, the physical universe and human existence within it. His vision is appealing; he unites the physical and the personal by means of the concepts of aim and value, which he sees as the keys to explaining traditional physical puzzles. Given the current popularity of theories of goal-oriented dynamical systems in biology and cognitive science, this approach

is timely. . . The most admirable aspect of this book is the willingness to confront every important aspect of human existence in the physical universe, and the recognition that in a complete explanation, all these aspects must be covered. Maxwell lays out the whole field, and thus provides a valuable map of the problem space that any philosopher must understand in order to resolve it in whole or in part."
Professor Natika Newton, *Philosophical Psychology*

"This is a very complex and rich book. Maxwell convincingly explains why we should and how we can overcome the 'unnatural' segregation of science and philosophy that is the legacy of analytic philosophy. His critique of standard empiricism and defence of aim-oriented empiricism are especially stimulating"
Professor Thomas Bittner, *Philosophical Books*

"I recommend reading The Human World in the Physical Universe ... for a number of reasons. First, [it] ... provides the best entrance to Maxwell's world of thought. Secondly, [it] contains a succinct but certainly not too-detailed overview of the various problems and positions in the currently flourishing philosophy of mind. Thirdly, it shows that despite the fact that many philosophers have declared Cartesian Dualism dead time and again, with some adjustments, the Cartesian view remains powerful and can compete effortlessly with other extant views."
Dr. F. A. Muller, *Studies in History and Philosophy of Modern Physics*

"Some philosophers like neat arguments that address small questions comprehensively. Maxwell's book is not for them. The Human World in the Physical Universe instead addresses big problems with broad brushstrokes."
Dr. Rachel Cooper, *Metascience*

"A solid work of original thinking."
Professor L. McHenry, *Choice*

Is Science Neurotic? (2004, Imperial College Press)

"This book is bursting with intellectual energy and ambition...[It] provides a good account of issues needing debate. In accessible language, Maxwell articulates many of today's key scientific and social issues...his methodical analysis of topics such as induction and unity, his historical perspective on the Enlightenment, his opinions on string theory and his identification of the most important problems of living are absorbing and insightful."
Clare McNiven, *Journal of Consciousness Studies*

"Is science neurotic? Yes, says Nicholas Maxwell, and the sooner we acknowledge it and understand the reasons why, the better it will be for academic inquiry generally and, indeed, for the whole of humankind. This is a bold claim ... But it is also realistic and deserves to be taken very seriously ... My summary in no way does justice to the strength and detail of Maxwell's well crafted arguments ... I found the book fascinating, stimulating and convincing ... after reading this book, I have come to see the profound importance of its central message."
Dr. Mathew Iredale, *The Philosopher's Magazine*

"... the title *Is Science Neurotic?* could be rewritten to read *Is Academe Neurotic?* since this book goes far beyond the science wars to condemn, in large, sweeping gestures, all of modern academic inquiry. The sweeping gestures are refreshing and exciting to read in the current climate of specialised, technical, philosophical writing. Stylistically, Maxwell writes like someone following Popper or Feyerabend, who understood the philosopher to be improving the World, rather than contributing to a small piece of one of many debates, each of which can be understood only by the small number of its participants.... In spite of this, the argument is complex, graceful, and its finer points are quite subtle.... The book's final chapter calls for nothing less than revolution in academia, including the very meaning of academic life and work, as well as a list of the nine most serious problems facing the contemporary world - problems which it is the task of academia to articulate, analyse, and attempt to solve. This chapter

sums up what the reader has felt all along: that this is not really a work of philosophy of science, but a work of 'Philosophy', which addresses 'Big Questions' and answers them without hesitation.... I enjoyed the book as a whole for its intelligence, courageous spirit, and refusal to participate in the specialisation and elitism of the current academic climate.... it is a book that can be enjoyed by any intelligent lay-reader. It is a good book to assign to students for these reasons, as well - it will get them thinking about questions like: What is science for? What is philosophy for? Why should we think? Why should we learn? How can academia contribute of the welfare of people? ... the feeling with which this book leaves the reader [is] that these are the questions in which philosophy is grounded and which it ought never to attempt to leave behind."
Margret Grebowicz, *Metascience*

"Maxwell's fundamental idea is so obvious that it has escaped notice. But acceptance of the idea requires nothing short of a complete revolution for the disciplines. Science should become more intellectually honest about its metaphysical presuppositions and its involvement in contributing to human value. Following this first step it cures itself of its irrational repressed aims and is empowered to progress to a more civilized world."
Professor Leemon McHenry, *Review of Metaphysics*

"Maxwell argues that the metaphysical assumptions underlying present-day scientific inquiry, referred to as standard empiricism or SE, have led to ominous irrationality. Hence the alarmingly provocative title; hence also-the argument carries this far-the sad state of the world today. Nor is Maxwell above invoking, as a parallel example to science's besetting "neurosis," the irrational behavior of Oedipus as Freud saw him: unintentionally yet intentionally slaying his father for love of his mother (Mother Earth?). Maxwell proposes replacing SE with his own metaphysical remedy, aim-oriented empiricism, or AOE. Since science does not acknowledge metaphysical presumptions and therefore disallows questioning them – they are, by definition, outside the realm of scientific investigation – Maxwell has experienced, over the 30-plus years of his professional life,

scholarly rejection, which perhaps explains his occasional shrill tone. But he is a passionate and, despite everything, optimistic idealist. Maxwell claims that AOE, if adopted, will help deal with major survival problems such as global warming, Third World poverty, and nuclear disarmament, and science itself will become wisdom-oriented rather than knowledge-oriented – a good thing. A large appendix, about a third of the book, fleshes the argument out in technical, epistemological terms. Summing Up: Recommended. General readers; graduate students; faculty."
Professor M. Schiff, *Choice*

"*Is Science Neurotic?* ... is a rare and refreshing text that convincingly argues for a new conception of scientific empiricism that demands a re-evaluation of what [science and philosophy] can contribute to one another and of what they, and all academia, can contribute to humanity... Is Science Neurotic? is primarily a philosophy of science text, but it is clear that Maxwell is also appealing to scientists. The clear and concise style of the text's four main chapters make them accessible to anyone even vaguely familiar with philosophical writing and physics... it is quite inspiring to read a sound critique of the fragmented state of academia and an appeal to academia to promote and contribute to social change."
Sarah Smellie, *Canadian Undergraduate Physics Journal*

"Maxwell's aspirations are extraordinarily and admirably ambitious. He intends to contribute towards articulating and bringing about a form of social progress that embodies rationality and wisdom... by raising the question of how to integrate science into wisdom-inquiry and constructing novel and challenging arguments in answer to it, Maxwell is drawing attention to issues that need urgent attention in the philosophy of science."
Professor Hugh Lacey, *Mind*

"Maxwell has written a very important book ... Maxwell eloquently discusses the astonishing advances and the terrifying realities of science without global wisdom. While science has brought forth significant advancements for society, it has also

unleashed the potential for annihilation. Wisdom is now, as he puts it, not a luxury but a necessity … Maxwell's book is first-rate. It demonstrates his erudition and devotion to his ideal of developing wisdom in students. Maxwell expertly discusses basic problems in our intellectual goals and methods of inquiry."
Professor Joseph Davidow, *Learning for Democracy*

"My judgement of this book is favourable...[Maxwell's] heart is in the right place, as he casts a friendly but highly critical eye on the Enlightenment Movement. 'We suffer, not from too much scientific rationality, but from not enough' he says...recommending a massive cooperation between science and the humanities...The book's style is refreshingly simple, clear".
Joseph Agassi, *Philosophy of Science*

"Nicholas Maxwell's book passionately embraces Francis Bacon's dictum that '[t]he true and legitimate goal of the sciences is to endow human life with new discoveries and resources'. The book's scope is commendable. It offers a thorough critique of the contemporary philosophy and practice of both natural (Chapters 1 & 2) and social science (Chapter 3), and suggests a remedy for what the author believes is the neurotic repression of the aforementioned Baconian aims."
Slobadan Perovic, *British Journal for the Philosophy of Science*

What's Wrong With Science? (1976, Bran's Head Books; 2nd ed., 2009, Pentire Press)

"Nicholas Maxwell believes that science (and also philosophy of science) should be humane and adventurous. In *What's Wrong With Science?* he boldly practises what he preaches. The argument is presented as a dialogue; and (rare among philosophers) Maxwell makes the debate lively and well-balanced...as a modern philosophical dialogue, the book is both instructive and great fun."
Jerry Ravetz, *New Scientist*

"how to be compassionate should be, on Maxwell's view, part of the process of our rational inquiry. Through compassion,

369

technology and applied science become humane. How this inquiry can be conducted with the desired result is by no means clear, but that its mastery is required for our survival in the technological age is certain."
Alan Drengson, *Philosophical Investigations*

"This is an unusual book...an unusually refreshing one."
T. A. Goudge, *Philosophy of Social Science*

'This rather peculiar and extremely provocative book.... is just "throwing open new possibilities, entertainingly indicating *Weltanschauung* that may not have occurred to people", and what a possibility he's (really entertainingly) opening! .... [T]he whole dialogue did work – the reading was pleasant and seemed almost real (rare events are those when philosophers actually do write in an attractive way)..... [I]t cannot be ignored that science has, at least in an indirect way, brought along not just prosperity but also grave global problems (global warming, arms of mass destruction, etc.). If our author is correct, which he probably is, that they are the "almost inevitable outcome" of science's failure to get rid of the philosophical idea of standard empiricism then he just might have a very good point. And even though this book, notwithstanding the hopes of our author, will probably not save the world, it will definitely not contribute to destroying it!'
Kristof K.P. Vanhoutte, *Metapsychology*

Science and the Pursuit of Wisdom: Studies in the Philosophy of Nicholas Maxwell (ed. L. McHenry, 2009, Ontos Verlag)

The name of Nicholas Maxwell for those who know him is tied to an original and revolutionary vision, and for those who do not know him it could be regarded as a token of a treasure to be discovered. Nicholas Maxwell is the man of his era; he observes the problems of his time and suggests a pervasive thought to solve them.
R. Ramezanivarzaneh, *Metapsychology*